Modern Educational Methods and Strategies in Teaching Mathematics

Modern Educational Methods and Strategies in Teaching Mathematics:

Changing Thoughts

By

Yousef Methkal Abd Algani

and Jamal Eshan

Cambridge
Scholars
Publishing

Modern Educational Methods and Strategies in Teaching Mathematics:
Changing Thoughts

By Yousef Methkal Abd Algani and Jamal Eshan

This book first published 2023

Cambridge Scholars Publishing

Lady Stephenson Library, Newcastle upon Tyne, NE6 2PA, UK

British Library Cataloguing in Publication Data
A catalogue record for this book is available from the British Library

ISBN (10): 1-5275-9112-3
ISBN (13): 978-1-5275-9112-7

TABLE OF CONTENTS

List of Figures and Tables

Figures

Tables

SUMMARY OF CONTENT

After a brief review of "Modern Educational Methods and Strategies in Teaching Mathematics: Changing Thoughts", the text covers topics as follows:

Chapter 1: Mathematics Education for School; Realistic Mathematics Education; Teacher Quality/Teaching Environment; Notation, Language, and Rigor; Assessment in Mathematics Education; Mathematical Modelling Approach in Mathematics Education; Research.

Chapter 2: Mathematics Education for Elementary School; The Five Building Blocks; Growth in early Elementary School; The Case of Elementary School Mathematics; Motivational and Affective determinants in Elementary School Mathematics; Difficulties in Learning Elementary Mathematical Concepts and Procedures; Conceptual and Procedural Knowledge; Developing Mathematical Knowledge for Teaching Elementary School Mathematics; Main sources of the Elementary Mathematics-Curriculum.

Chapter 3: Influence of Mathematics on Secondary Students; Learning preferences and Mathematics Achievement of Secondary Students; The Construction of Identity in Secondary Mathematics Education; Transition from Primary to Secondary School Mathematics; Active Learning in Secondary School Mathematics; Self-Concept, Self-Efficacy, and Performance in Mathematics; Perceptions of Mathematics Education Quality on Mathematics Achievement.

Chapter 4: Outlooks of Senior Secondary Students on Mathematics Learning; Increasing Senior Secondary School Mathematics Achievement; Problems and Interventions in Senior High School Mathematics; The Reasons for Poor Math Performance in Senior Secondary School Students; Approaches for Improving Mathematics Performance in Senior Secondary School Students; Perceptions of Difficult Concepts in Senior Secondary School Mathematics Curriculum; Self and Cooperative-Instructional strategies for Improving Senior; Secondary School Students' Attitudes toward Mathematics.

Chapter 5: Curriculum and Pedagogy in Mathematics; Curriculum Design; The Processes of Planning and Development of Curriculum; Curriculum Development; Models of Curriculum Development; Evaluation of Curriculum; Pedagogy of Mathematics; Pedagogy of Mathematics Teaching.

Chapter 6: Teaching and Learning Maths in School; Students Engagement in Learning Maths; Teachers' Beliefs about Teaching and Learning Mathematics; Using Technologies in Teaching and Learning Mathematics; Elementary Students Learning Maths; Mathematical Problem-Solving of Middle School Students with Learning Disabilities; Gender Differences in Perception of Teaching and Learning Mathematics.

Chapter 7: Innovation in Mathematics Education; General Objectives of Teaching Mathematics; The Innovations in Mathematics Education Project; Reformations in Mathematics Education over the years; Need of Innovative practices in Mathematics Education; Innovative methods in Mathematics Education in school; Mathematics learning strategies for the 21st century learners.

Chapter 8: Cognitive Development; The Individual Versus the Environment; Major theories; How Do Children Learn?; The Brain and Cognitive Development; Educational implications; Piaget's Theory of Cognitive Development.

Chapter 9: Cognitive Neuroscience in Mathematics; Cognitive Neuroscience Meets Mathematics Education; Developmental Cognitive Neuroscience of Arithmetic; Cognitive Neuroscience Contribution in Mathematics Education; Cognitive Neuroscience and Teaching; Cognitive Neuroscience and Mathematical Reasoning; Numeracy and Mathematical Learning Cognitive Neuroscience.

Chapter 10: Issues and Challenges; Basic Challenges and issues in Mathematics Education; Difficulties in Mathematics; Teaching Practices; Issues in Mathematics; Issues in Elementary Education; Issues in Secondary Education; Issues in Senior Secondary Education; Challenges in Mathematics Education; Summary.

PREFACE

Mathematics is the peak of precision, as evidenced by symbols in calculations and formal proofs. Symbols, after all, are just that: symbols, not ideas. Mathematics' intellectual content is found in its ideas rather than the symbols themselves. In other words, the intellectual content of mathematics is not found in the symbols, where the mathematical rigor is most immediately visible. Human thoughts, instead, are the source of the problem. However, mathematics does not and cannot scientifically analyze human concepts on its own; human cognition is not its domain. It is up to cognitive science and neuroscience to achieve what mathematics cannot: apply the science of mind to human mathematical concepts. The human brain and mental capabilities constrain and structure mathematics as we know it. The only mathematics we know or can know is based on the brain and thinking.

Traditional school mathematics is typically defined by an overreliance on meaningless paper-and-pencil computations, precisely written problems that follow prescribed methods, and a focus on meaningless symbolic manipulation. The importance of individual seatwork is highlighted. The authority sources for determining the validity of an answer are the teacher, the textbook, and the answer key. Many of these traits of traditional classroom mathematics still exist, but they are competing with other qualities as teachers attempt to modify the way they teach mathematics. When we enter a math classroom these days, we are likely to observe kids working in groups, manipulating materials, and talking and writing about mathematics. They frequently engage in mathematically demanding tasks, and the textbook (if used at all) serves as yet another resource in the classroom.

However, there can be significant variances between classrooms. Even reform-oriented classrooms are part of a larger framework in which different perceptions of what constitutes mathematics influence what happens in them. Mathematicians frequently refer to a specific element of playfulness in their work of "messing around" with ideas in their search for justifications, counterexamples, and so forth; mathematicians often refer to particular element of playfulness in their work, of "messing around" with ideas in their search for justifications, counterexamples, and so forth;

mathematicians often refer to a particular element of playfulness in their work, of “messing.

This book looks into the subject of classroom mathematics education. Students’ understanding and enthusiasm for mathematics grow as they progress through elementary school, as do their thinking skills. The five building blocks of learning mathematics for primary children and motivational and affective determinants in elementary school mathematics were defined here. Mathematical aptitude is also critical for a society’s economic success. Other professions, such as engineering, sciences, social sciences, and even the arts, require a firm grasp of mathematics. Thus, identity formation and how students prefer to learn are stated in secondary mathematics education.

In general, humans use mathematics daily, and this daily use of mathematics causes the human brain to express issues, ideas, and solutions necessary for the human race’s survival, so this book explains how to teach math to Senior Secondary School students using problems and interventions. The book also included a quick overview of curriculum design and the main components of curriculum development. It will be shown in the following sections that mathematics education and learning may be viewed as a progressive system. It also included student involvement in learning and teacher ideas about teaching mathematics, the use of technology in math through cellphones and gaming, and the implications of developmental cognitive neuroscience for learning and teaching for academic success. The book came to a close with a discussion of the challenges and benefits of the mathematical education system.

CHAPTER 1

MATHEMATICS EDUCATION FOR SCHOOL

1.1 Introduction

Amount (numbers), organization, space, and change are all studied in mathematics. Mathematicians and philosophers have differing perspectives on the scope and definition of mathematics. Investing in children's teaching reflects a nation's sense of modernity and progress. If science education has frequently been referred to as a social asset in the imagined prospect, learning in the "great roads of math" may be its desire for the yet-to-be-envisioned future (National Board of Higher Mathematics, 2012). Presidents and Prime Ministers emphasize the importance of mathematics and science education in preparing the nation's children to face the tests of the 'modern economy. Many countries recognize the importance of developing a mathematically literate populace and envision a powerful mathematical elite capable of shaping the knowledge-based society of the twenty-first century. Simultaneously, mathematical ability is universally seen as complex to attain (NCERT-National Council of Educational Research and Training, 2006).

In opposition to the aspirations of the globalist elite, the student's goals and aims must be addressed. Education can be seen as the essential weapon for breaking out of poverty in a mainly impoverished population (by any criterion). The capacity to 'calculate,' 'evaluate,' and 'predicts' are essential life skills that education must (and ideally does) teach. One feels the intelligence of dissatisfaction that such talents are not imparted through a formal curriculum. A grocer passionately protested in 2006 at a public hearing where a curriculum committee met members of the public whom he could never attract cultured young people who could estimate whenever inventories needed to be refilled and by how much (National Board of Higher Mathematics, 2012).

Thus, how would you describe math education in school? We believe this was a combination of severe systemic issues. However, growing young people face them with optimism, in a place with numerous ideas, creativities, and a chaotically running system.

1.2 Realistic mathematics education

RME refers to Realistic Mathematics Education, a domain-specific math teaching concept developed in the Netherlands. RME is distinguished by the prominence of rich, "realistic" scenarios in the learning process. These scenarios form a basis for creating math models, methods, and methods, as well as a framework in which students can subsequently use their knowledge of mathematics, which became more conventional and generic over time and low context detailed. While 'realistic' circumstances in the sense of 'real-world' situations were crucial in RME, the term "realistic" here has a broader definition (Van den Heuvel-Panhuizen, 1996). It suggests that students are presented with problem scenarios that they could imagine. The definition of "realistic" comes from the Dutch phrase 'zich realiseren,' which means "to imagine." RME gets its name from this emphasis on creating everything real in your imagination. As a result, challenges taught to students in RME could come from the actual world and the fantasy land of the formal world of maths, as long as the issues were contextually actual in the students' minds (Greer & Brian, 2009).

1.2.1 Evolution

Mathematical evolution can be viewed as an already set of abstractions or as a broadening of subject matter. It is recognized that, for example, a gathering of 2 apples and a group of two oranges have something similar, namely the number of its individuals. It was possibly the first concept accepted by many creatures. In adding to counting physical objects, as shown by bone tallies, prehistoric populates may have recognized how to tally notional amounts such as time, days, season, and years. Further advanced mathematics did not even exist until the Egyptians and Babylonians started with geometry, algebra, and arithmetic for taxation and other economic computation, engineering and astronomy, and structure, approximately 3000 BC. Trading, land surveying, painting and weaving designs, and time recording were among the first applications of math. Primary arithmetic (subtraction, addition, division, and multiplying) first occurs in the archaeological evidence in Babylonian mathematics. Maths preceded writing, and number systems were numerous and varied, with Egyptians creating the earliest known written numerals in Middle Kingdom documents, including the Rhind Mathematics Papyrus (Gravemeijer &

Koeno, 1994). With Greek mathematics, the Greeks and Romans carried out, between 600 and 300 BC, a scientific study of math in its rights. Since then, mathematics has grown significantly, and there has been a beneficial interplay between mathematics and science, which has benefited both. Mathematical breakthroughs are still being made today. "The percentage of books and papers shown in the Math Review sites database since 1940 was now more than 1.9 million, and more than 75 thousand substances were decided to add to the file every year" (Mikhail B. Sevryuk) in the January 2006 issue of the Bulletin of the American Arithmetic Society. New mathematical theories and proof are found in the vast majority of publications in this ocean (Van den Heuvel-Panhuizen et al., 2020).

1.2.2 Inspiration, pure and applied mathematics, and aesthetics

Math is derived from a wide range of issues. These were first discovered in trade, land measuring, architecture, and astronomy; now, all disciplines propose difficulties that mathematicians should investigate, and many of these issues originate inside mathematics itself. For instance, to use a mixture of math skills and physiological insight, the physicist Richard Feynman created the integral path formula of quantum mechanics. Today modern string concept, a still-developing technical hypothesis that tries to unify the four fundamental natural forces, endures outstanding new mathematics (d'Ambrosio & Ubiratan, 1986).

Specific mathematics is only helpful in the field in which it was developed and is only used to address issues in that field. However, mathematics influenced by one domain frequently proves helpful in several areas and becomes part of the overall stock of mathematical conceptions. The contrast between pure mathematics and applied mathematics is frequently made. Nevertheless, pure mathematics concepts, such as number theory in cryptography, frequently have used. As Eugene Wigner put it, the "unreasonable efficiency of mathematics" is an extraordinary reality that even the "purest" arithmetic frequently finds practical uses. The growth of information in the modern period has led to emphasis, as it has in many fields of study. There are currently 100s of emphasized topics in mathematics, and one of the topics is Mathematics Taxonomy with 46 pages long. Specific fields of applied mathematics, such as stats, operations research, and computer science, have fused with relevant societies outside of mathematics to become fields in their own right (Skovsmose & Ole, 1990).

Most mathematics has distinct effects observed for technologically proficient people. Most mathematicians speak about mathematics' refinement, natural aesthetics, and natural self. The importance of simplicity and generalization is emphasized. A straightforward and graceful demonstration, including Euclid's proof of which there are substantially many significant factors, and an attractive arithmetical approach, including the quick Fourier transform, are both beautiful. In a Mathematician's Apology, G.H. Hardy articulated the idea that these aesthetic factors were adequate to warrant the investigation of pure mathematics on their own. He highlighted elements that add to a mathematical beauty: relevance, unpredictability, inevitability, and economy. Mathematicians frequently seek out extremely elegant arguments, such as proofs from 'The Book' of God, by Paul Erdos. Another measure of how many people like solving mathematical problems is the prevalence of recreational mathematics (Ernest & Paul, 1985).

1.2.3 Structure

As illustrated in Figure 1.1, the education system is divided into periods of development ranging from pre-primary to post-graduate. Elementary (primary and upper elementary) and secondary education were governed separately. Undergraduate education lasts three years, whereas professional degrees last 3-4 years. Universities were controlled centrally but administered locally, with a system of associated colleges that provide undergrad degrees (Halmos & Paul, 1981). The whole education system is governed by the Ministry of Human Resource Development, with every state government holding its own Education Ministry and a Central Advisory Board of Education serving as a stage for interactions between the Centre and states. In total, 43 Boards of School Education exist across the nation, and they are responsible for developing curricula, training teachers, and granting certification. The National Council of Educational Research and Training oversees school education (Keitel & Christine, 1989).

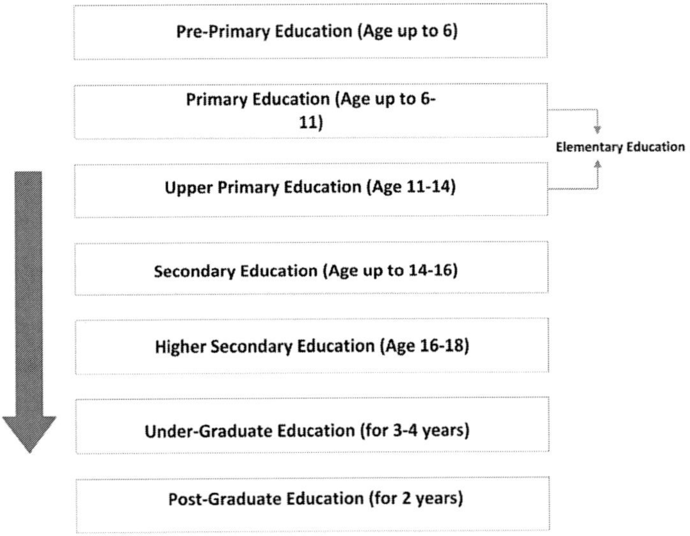

Figure 1.1 Different levels of education

The Ministry of education is the leading organization for syllabus concerns, except for the Central Board of Secondary Education, which determines syllabi. Each university develops its specific curriculum, but the University Grants Commission oversees how they are implemented. A thriving Open University system and the National Institute of Open Schooling aim to give access to education by bridging possible gaps created by these systems (Bishop & Alan, 1988).

1.3 Teacher Quality/Teaching Environment

The educational environment is dismal for the majority of students. The school comprises a one-room schoolhouse with one teacher covering various grades and 30 students per educator. It is worth noting that many remote public schools lack even the most necessities. Aside from these challenges, many rural schools were severely understaffed by a lack of resources. Though one teacher might well have forty students in a class on the median, the refusal of many instructors to take remote rural postings (attempts to change these posts through lobbying and court disputes) means. However, the enrolled student ratio was significantly higher in many rural areas. High levels of teacher absence and low levels of teaching activity worsen high student-teacher ratios (Belsito & Courtney, 2016).

Primary education is not easy to scale up. It enhances instructional results for young students unless it makes significant determinations to recruit extensive statistics of new teachers and invests properties in upgrading school infrastructure. Education is a well-paid profession, and most teachers were hired primarily on their political ties rather than their subject or pedagogical competence. There was no mechanism to encourage instructors to progress in their academic performance, and inadequate training was obtainable to assist teachers in improving their teaching methods. Whereas the 1986 National Policy on Education improved the required curriculum by raising English and science standards, the revisions were not accompanied by a new instruction and assessment methods. The preceding is a synopsis of a typical school day from the British Council. Although this statement is undoubtedly accurate for urban private schools, this is not the situation in the ordinary government school, as rote learning is still the central objective for passing exams: A typical school day starts between 8:00 and 8:20 a.m. Students go to school for around 6 hours each day, with 40-50 minutes classroom periods and a half–hour lunchtime. Some schools operate for an extended time (about 8 hours) with two breaks. Typically, the school day begins with a morning assembly that incorporates prayers, meditation, important news, special assemblies, Mass Exercises, or any other organized activity for all students and instructors (Moore & Robert, 1994).

Primary classes are often shorter than secondary classes in terms of time. A teacher is in charge of the grade, and another instructor assists her. This is not always the case, and in some remote areas of the country, a single instructor may be in charge of several classes31! In a secondary school, there is usually one class teacher in charge of the entire grade and teaches one or more topics. Throughout their separate sessions, subject professors instruct pupils. Students were obliged to change to the appropriate work areas in the school for several topics such as Arts, Physical Education, Music, Work Experience, Library, Science Practicals, etc. A class in a city usually contains rows of benches and a teacher's desk in the front. One or more bulletin boards (blackboards)/ exhibition spaces and one/more store cabinets are other essential classroom tools. Private schools in cities may have more sources to set up current classrooms, but many are working to develop Hi-tech schools with computer networks and sophisticated coaching tools. On the other hand, some schools in rural and semi-urban areas lack even the most basic amenities such as concrete structures, seats, and chairs.

Nevertheless, many projects are underway in the field of education, particularly to reach out to impoverished societies. A regular classroom session entails the teacher providing directed instruction based on the set curriculum. Teachers were progressively employing the Project style of instruction to motivate pupils to think critically, conduct research, and compile data. The bulk of the work and evaluation is done individually, but students do group work on specific topics and projects.

1.4 Notation, language, and rigor

It was not until the 16th century that most of today's mathematical logic was devised. Math was formerly written out in words, a time-consuming technique that hampered mathematical detection. Many of the notations being used now were created by Euler (1707–1783). Current writing simplifies mathematics for experts, which cannot be exceptionally comforting for newbies. It is incredibly condensed: just some symbols hold a countless deal of information. Applied mathematics notation, like musical notation, has a precise grammar (that varies to some range between authors and disciplines) and encodes data that would be hard to convey in any other way (Harel et al., 1996).

For beginners, mathematics might be hard to decipher. The words like or and only are more specific than in everyday speech. Furthermore, words like open and field have been assigned mathematical definitions. In mathematics, words like homeomorphism and integrable have specific definitions. In addition, mathematical jargon includes abbreviations such as iff for "if and only if." Mathematics needs more accuracy than everyday speech, which is why special notation and technical language are used. The precision of language and reasoning is referred to as "rigor" by mathematicians.

Proof in mathematics is primarily an issue of rigor. Mathematicians desire their findings to follow its axioms systematically. It is to prevent erroneous "theorems" based on flawed intuitions that have happened frequently throughout the field's history. The level of rigor needed in mathematics has changed over time: the Greeks demanded thorough explanations, but the techniques used by Isaac Newton were less rigorous. In the nineteenth century, difficulties with Newton's concepts would lead to a renaissance of meticulous study and formal evidence. Many of the most popular misconceptions about mathematics stem from a lack of rigorous knowledge. Computer-assisted theorems are still a source of contention between mathematicians today. Due to the difficulty of verifying extensive computations. It is possible that such proofs are not stringent enough.

Axioms are formerly supposed to be "self-evident truths"; however, this notion is flawed. An axiom was essentially a string of symbols on a comprehensive level, with intrinsic value only in the context of all derivable formulae in an axiomatic system. Hilbert's program aimed to axiomatize all of mathematics. However, according to Gödel's incompleteness concept, every axiomatic system has undecidable formulae, making a comprehensive axiomatization mathematically impossible. However, mathematics was frequently thought to be nothing more than set theory in some axiomatization, so any mathematical assertion or proof might be put into equations within formal logic (Niss & Mogens, 1993).

1.5 Assessment in Mathematics Education

Classroom educators have long used varied assessment methods to observe their students' learning abilities and inform future guidance. On the other hand, policymakers worldwide constantly use external evaluations to gauge a country's students' math skills and, in some cases, to compare a specific understanding to that of students worldwide. As a result, external examinations frequently impact classroom teachers' teaching techniques. Evaluation is a subject of interest to educators at all levels due to the emphasis placed on it by many stakeholders (Ginsburg & Herbert, 2009).

The Co-chairs and committee members of the two Topic Study Sessions (MITZAV) at RAMA decided to focus on evaluation in mathematics education: RAMA: The National Authority for Measurement and Evaluation in Education in Israel. Formative Evaluation for Mathematics Learning has chosen to work on developing this volume, to recognize discussions and research which may overlap and those that are exact to either class or large-scale valuation in mathematics education. We wish to include an everyday basis for conversations by creating this field questionnaire and evaluating the progress made in evaluation around the world.

This book uses research to address these concerns. It emphasizes particular distinctions concerning the challenges, difficulties, limits, and affordances that come with large-scale and class evaluation in mathematics education, along with some similarities. We acknowledge that the traditions, procedures, intentions, and issues of these two evaluation forms differed vastly. However, we believe there are some areas where there are overlaps that should be discussed, including such assessment item and task design, as well as links and consequences for professional education and experience.

1.6 Mathematical Modeling Approach in Mathematics Education

Rapid progress in technology and information has altered expectations of society regarding individuals and schools. In today's environment, mathematics instructors were intended to increase the amount of students who can devise practical solutions to real-world issues and apply mathematics efficiently in their everyday lives (Saxena et al., 2016). As a result, instead of being afraid of mathematics, kids will like it and understand and appreciate its relevance and power. This procedure of growth and advancement prompted new inquiries in our education systems, which became necessary to test novel educational techniques, methods, and models. Model-based instruction is one of the emerging methods of mathematics education. According to Pesen (2008), models are concrete beings, visuals, and objects that reflect some states of an idea that need to be built. This method has piqued the curiosity of students. The fundamental reason that math is the most significant global education field is that it can be applied in a variety of ways in fields and topics which are unrelated to it. Outside of mathematics, it is frequently employed in a veiled or overt manner, particularly in circumstances of difficulties, situations, or regions involving mathematical models and modeling. Mathematics, according to Freudenthal, was not a closed environment or a subject that should be studied but rather human impacts that must have a connection to reality.

When reviewing the published papers in mathematical modeling, it is clear that much research has been carried out on this topic in other countries. Nevertheless, there has been little research in Turkey on applying the modeling technique in math education. Furthermore, our studies lack a precise definition of mathematical modeling and modeling. The word mathematical modeling is defined in this study, and good examples of models that teacher educators and aspiring mathematics teachers could use in their teaching methods were presented as instances of models found in foreign literature. This research looks at mathematical modeling from a theoretical standpoint and in-class applications in Turkish primary, secondary, and high schools.

Students who study social constructivism build and reorganize their knowledge. To do so, students should participate in problem-solving and exploration tasks, as well as conversations with peers and teachers and experiences to convey their ideas in various ways. Education, according to constructivism, does not occur through the transfer of knowledge but rather through the process of asking questions, researching, and solving issues.

The constructivist method and many general intelligence concepts are at the heart of the new educational programs established by the Ministry of National Education (MNE) in 2005 (Math, Turkish, and Science Technology courses and Social Studies). As a result, active learning strategies and approaches are used in the classroom. The variation in the teaching process is the most critical difference between conventional and constructivist approaches. There has been a shift of focus from a teacher-centered strategy to a student-centered approach. As a result, group work, role play, act-out actions, games, and exploration actions are being used in learner-centered classes in general, and practical mathematical effective teaching, guided finding strategy, math games, and mathematical process model have been used in mathematics in specific (Wigley, 2008).

1.6.1 The major challenge

If someone was to pick one obstacle out of the issues we have discussed as the most essential, it might have to be the creation of a pool of qualified mathematics educators in sufficient numbers. The numbers are there at the primary level, but not with the necessary grasp of mathematics, dispositions into arithmetic, or knowledge of how students learn (fail to study) maths. The societal disparities and resource-poor rural schools necessitate more teacher capability than in more prosperous, democratic cultures. This necessitates the development of new models of teacher development that have yet to be defined. The numbers can be intimidating at higher levels. The current teacher pool is severely insufficient to satisfy the demands, particularly when universal school education becomes a plausible reality within a century. The difficulty of rigor and breadth in mathematical education and experience becomes more severe when dealing with numbers. Developing comprehensive strategies to improve teacher preparation excellence is likely the most pressing need in the math landscape (Freudenthal and Hans, 1981).

1.7 Research

Studying math education is an essential agenda item for mathematics education. Because of its traditional structure, academic institutions that do educational research tend to draw many persons who are neither qualified nor interested. Furthermore, the concept of research delivering answers to curricular conundrums or pedagogy is still outside the realm of educational judgment, which is not to minimize the significant contributions made to reform by both government and non-government projects marked by ingenuity and commitment (English et al., 2012). These, though, do not yet stand on a foundation of inquiry and scathing criticism. The system must develop a method for conducting research on a few fronts in response to well-formulated queries and using the results to affect policy. Internal critique of the field of mathematics and its education and practices are on the agenda for such inquiry. The civilization and its cultural and professional habits provide possibilities for mathematical investigations that a pedagogue could include in their toolset. Nevertheless, a body of research must be established to make practical use of such prospects.

1.8 Summary

Mathematics is increasingly structuring reality, and the ordinary person is becoming increasingly free of the need to apply mathematics. There is an incalculable amount of mathematical knowledge available now, and it is continually developing, and there are people who utilize specific areas of it professionally. However, its extent and eventual specialization place it outside of public education. This chapter discussed realistic mathematical education, the quality of teachers, and the educational environment. This chapter also goes over the mathematical modeling approach in mathematics education.

CHAPTER 2

MATHEMATICS EDUCATION FOR ELEMENTARY SCHOOL

2.1 Introduction

Mathematics is considered the foundation for the growth of some other regions of science, and it is required in elementary education. In general, pupils find maths difficult. As a result, primary school math instruction should be enjoyable and engaging for pupils. Since these pupils might have all job opportunities available to students, early elementary school maths is the same for all learners (Ariani et al. 2017). The objective is to guarantee that all pupils have equal access to opportunities. The essential basic characteristics of mathematics are precision, lack of ambiguity and concealed assumptions, and mathematical reasoning, which are difficult to overstate their significance. A subject is not a maths problem unless established and has a distinct set of answers. In a real mathematics problem, it simply cannot be any concealed assumptions. Term definitions, procedures, and representations must all be precise. Though it is simple to state that maths is rational, describing mathematical thinking is more challenging. The framework of mathematics is built on mathematical reasoning. As a result, a learner should assimilate the reasoning skills that produce a mathematical notion to comprehend it.

Inverting and multiplying decimals is an essential mathematical ability, but mathematical reasoning illustrates why inverting and multiplying fractions is the proper process. This type of mathematical reasoning must be explained, and technical capability is learned using mathematical reasoning. It is a crucial concept that will be considered throughout the debate. Without knowledge, talents are useless, and understanding without talent is useless. Elementary mathematics is made up of five fundamental building blocks. Keep in mind that maths is precise everywhere. There seem to be no concealed implications or confusing assertions. Definitions are necessary and should be accurate. It is all held together by rational thinking, and mathematics permits everyone to solve problems. The fundamental building blocks are not only the framework on which math is built but also prepare students for algebra and mathematics beyond algebra if taught correctly.

2.2 The Five Building Blocks

2.2.1 Numbers

Mathematics is built on the foundation of numbers. Thus children should know how to count and have fast memory of single-digit number concepts for arithmetic operations (and the related facts for subtraction and division). Quick memory enables the students to focus on learning various ideas and solving problems. It is essential in advanced mathematics

2.2.2 Place Value System

The place value system is a complex mechanism for effectively representing whole numbers. The five key construction elements are organized and unified by such a philosophy. However, its significance is frequently neglected; it is the core of the numerical system and, as such, needs far more consideration than it now receives (Wilson 2009). There is much more to it than hundreds, tens, and ones. The basis for university maths is arithmetic and algebra. The cornerstone for both arithmetic and algebra is knowledge and learning of the place value system and how it is applied. Except in the framework of the place value system, mathematical procedures can be explained. Because comprehension is so important, It will start with the place value system. Students must be prepared for algebra in elementary school. Working with equations is only a modest extension of the place value system in algebra. The place value system is an integral part of algebra practice.

2.2.3 Whole Number Operations

In addition, subtraction, multiplication, and division of whole numbers represent the basic operations of mathematics. Much of mathematics is a generalization of these operations and rests on an understanding of these procedures. They must be learned with fluency using standard algorithms. The standard algorithms are among the few deep mathematical theorems that can be taught to elementary school students. They give students power over numbers and, by learning them, give students and teachers a common language. The basic mathematical operations are addition, subtraction, multiplication, and division of whole numbers. Most of the maths is a derivation of such techniques, and it is based on a comprehension of them. It should be learned utilizing established methods and with ease. Standard algorithms are one of the few fundamental mathematical theorems which can be explained to pupils in primary school. They provide learners control over numerals and, by understanding them, numerals provide a common language for students and teachers.

2.2.4 Fractions and Decimals

The four primary mathematical operations using whole numbers should be applied to fractions and decimals and be viewed as an extension of whole numbers. If students want to continue further mathematics, they need to become competent in such calculations for fractions and decimals. Learning fractions is essential for algebra preparation. Knowing ratios in various settings, especially commerce, requires firm foundation infractions.

2.2.5 Problem Solving

Single, two, and multi-step problems (those requiring so many processes to solve) must be covered through a student's mathematics instruction, particularly word or narrative questions. Every new concept or skill should be applied to a sequence of more complex tasks. Critical thinking skills include the ability to translate words into maths and the ability to solve multi-step problems. The purpose of mathematical education is to improve critical thinking.

2.3 Growth in early Elementary School

Researchers looked at performance development throughout third grade of pupils who started preschool with low, medium-low, medium-high, and high mathematics skills, as well as the influence of educator time on mathematics teaching and student involvement in such progress. It was discovered that pupils who started with a minor success level also showed the lowest improvement over time. Students in the second top skill groupings grew at equal rates and the same rates. Pupils in the lowest group had the most fantastic time in class, yet they were the least engaged (Bodovski and Farkas 2007). All pupils improved their grades when they spent more class time, but the effect of involvement was most prominent among the least performing students. The lowest-performing group's reduced involvement accounted for more than half of the poorer performance gain in grade three. Teachers must make more significant efforts to increase the group's basic knowledge and academic participation if inequity in mathematics performance is to be lessened. Because it has already demonstrated a significant learning rate, researchers predicted the top learners to be more engaged with classroom education.

Furthermore, there is still the chance that their extraordinary performance will bore them. There is expected to be considerable financial and racial discrimination by a school, considering residential segregation trends by social class and race and the use of local primary schools. Educators might offer additional teaching time to high-achieving students due to grouping, based on the notion that "students are willing to learn, and the parents would like them to be engaged." Conversely, since these kids "learn fast, are already ahead, and it is too soon to start teaching the material required for the next best grades," educators can devote less teaching time. There are also conflicting alternatives for these children's growth paths. Researchers projected that the children with a strong foundation in mathematics are affluent and will receive high-quality instruction; their overall level of growth would be the fastest. Regression to the average and maximum impact in instruction and training, on the other hand, may serve to limit its increase.

The improvement in preschool was calculated by subtracting the autumn preschool score from the spring kindergarten score. Researchers calculated the first-grade progress by subtracting the springtime preschool score from the springtime first-grade result. Finally, the springtime third-grade result was subtracted from the spring first-grade mark to determine the final performance result. It is vital to note that children who start with the lowest mathematical skills also demonstrate the lowest growth. It was also discovered that basic numerical understanding and skills significantly impacted children's development within this least-performing group. There is much variation in pupils' mathematical ability in the autumn of preschool, which goes against the popular idea that mathematics learning starts in school, not in preschool. Compensatory maths training in the early years looks essential, especially for underprivileged pupils. The emphasis on mathematical fundamentals must complement a significant emphasis on reading fundamentals for the age group.

2.4 The Case of Elementary School Mathematics

Researchers aimed to determine if teacher content practices are influenced by teachers' professional attitudes and values or by school policies and other external factors. Distinguishing between excellent and terrible is dangerous, as it necessitates judgments beyond the evidence. There is a risk in the hopes of developing new and more profound knowledge of the shortcomings of what is taught in elementary school mathematics, as well as insights into how those shortcomings might be addressed (Porter 1989).

2.4.1 Emphasizing Problem Solving

Educators spend approximately half of the time not focused on skill development and conceptual understanding and the other half teaching is problem-solving abilities, primarily narrative problems. In these assessments, applications are defined as activities in which the methods required to solve a problem were not stated clearly but assumed in the problem's representation. Although if learners studied on a page in which every narrative issue included the addition of only one value, all the work on problems was considered as applications. However, at some point during the completion of such a task, the uncertainty of the answer should vanish, and the remaining difficulties will resemble skill practice rather than problem-solving. As a result, the conclusion that fourth-grade instructors spent just 11% of their maths teaching time on applications/problem solving should be seen as an underestimate of the actual amount.

The importance of skill development educators place on their students is represented in the textbooks they utilize. According to content evaluations of fourth-grade books, 65 percent to 80 percent of the activities focused on skill practice, 10% to 24% on conceptual comprehension, and 6% to 13% on problem-solving. However, if teachers want to change their focus away from skills and toward problem-solving and cognitive thought, they will need guidance to help them do so. As a result, teacher content practices are entirely compatible with poor student performance results when assessing the importance of solving problems and higher-order thinking in primary school mathematics. Problem-solving is rarely taught, and as a result, student achievement is low.

2.4.2 Teaching for Exposure

The second aspect of elementary school mathematics training that is almost as problematic as the lack of emphasis on problem-solving and conceptual knowledge is the absence of emphasis on problem-solving and conceptual understanding. A vast proportion of the subjects covered in class are briefly covered, if at all. To put it differently, only 20% to 30% of what teachers presented received the same amount of attention throughout the year as one brief thirty-minute lesson; below 10% of the topics got two hours or more of teaching. The startling discovery contradicts to focus on problem-solving and critical thinking.

It is challenging to characterize the education technique for exposure in elementary school mathematics. Educators, on the one hand, are unrepentant about the activity. On the other hand, educating for exposure is unlikely to improve student performance and is likely to transmit undesired attitudes to students. Learners may assume that conceptual knowledge and application are less significant than efficiency and agility in computing skills since a greater proportion of topics dealing with conceptual knowledge and implementation are offered for exposure than topics dealing with skills.

2.4.3 A Slow-Moving Curriculum

The pace with which curriculum varies as pupils advance through the levels is a third distinguishing aspect of primary school mathematics. Separating elementary school mathematics into grades is, without a doubt, random. Subjects initiated just after one level are extended into the start of the next grade, resulting in some overlapping content between grades. To a certain level, issues are reviewed over and over to gain a more profound knowledge every time. Concerns about topic overlap among grades, like those involving problem solving and teaching for mastery, are matters of degree. The research of fourth and fifth-grade teachers in 17 schools comprehensively assesses the discrepancy between fourth and fifth-grade students by adjusting for between-school variation.

Even though the overall conclusion would be that the fifth-grade subject is too similar to the fourth-grade subject, these grade-level differences yield some favorable outcomes. From fourth to fifth grade, the majority of time invested in teaching skills decreased significantly (from 76 percent to 70 percent), whereas the proportion of time invested in educating application/problem solving improved from around 10% to approximately 20%. The amount of time spent acquiring conceptual knowledge remains relatively consistent. The finding, combined with the fact that fifth-grade educators spend nearly ten hours teaching mathematics (8,485 minutes) than fourth-grade teachers (7,860 minutes), implies that fifth-graders obtain more than double the amount of problem-solving training as fourth graders.

The above analysis suggests the following recommendations for change:

- Educators in elementary schools should emphasize conceptual knowledge and give opportunities for students to use concepts and skills in constructing and resolving mathematical problems.
- A smaller number of topics must be discussed in greater depth. All effort is being made to establish that the assumption that the contents are given will be understood. It should increase overall academic performance and views about the topic.
- The mathematics curriculum must be integrated appropriately among grades so that what is learned a year is not repeated the following year. It should also encourage pupils to take mathematics earnestly. Currently, the benefit of understanding anything before it is even taught is boredom from unnecessary repeating of it over time.
- Mathematics must be introduced at a set time each day. Additional tasks should barely interrupt or pre-empt that period.

2.5 Motivational and Affective determinants in Elementary School Mathematics

According to some researchers, strategic planning is a time-consuming and effortful activity; as a result, pupils who appreciate an assignment or domain will be more likely to be using strategy to enhance their chances of success, such as cognitive and metacognitive strategies that require more mental focus, dedication, and consciousness, and also greater motivation level. Research has also found that task value is a significant factor in students' adoption of cognitive and regulating strategies. Pupils in seventh and eighth grades who appreciated and were engaged in the academic areas and content were more inclined to use more excellent methods and personality techniques. In the case of maths, new data imply that self-regulation and cognitive strategy utilization (i.e., rehearsal, organization, and evaluation procedures) are linked positively with the importance of pupils' place on the subject (Chatzistamatiou et al. 2015).

Moreover, studies and theories imply that emotions experienced in learning environments could have a significant and long-term effect on pupils' personalities and success. Positive attitudes, such as excitement, appear to be highly linked to deeper cognitive processing (clarification and regulation), critical thinking, and metacognitive monitoring mechanisms in pupils. Several researchers have found that pupils' stated enjoyment of mathematics learning is linked to their ability to self-regulate their learning and their usage of strategies. Improvements in positive emotions in seventh-graders, such as joy and confidence, were consistently linked to changes in personality methods and math skills. A fall in joy, for example, was linked to a drop in self-regulated learning processes.

2.6 Difficulties in Learning Elementary Mathematical Concepts and Procedures

Students can develop what they already know to learn some new mathematical concepts and methods. In other words, studying with comprehension entails building connections or offering solutions within or between current knowledge and innovative information. Most mathematical concepts, such as numerals, measures, and probability, are intuitively understood by pupils before they enter school. For instance, by performing out problems involving connecting, splitting, or matching numbers with collections of objects, kindergarten and first-grade pupils naturally solve a range of problems involving connecting, splitting, or matching quantities. Variations of these principles can be used to help students learn about addition, subtraction, multiplication, and division (Yetkin 2003).

Students have trouble studying elementary mathematics regardless of their understanding of certain maths concepts since they are typically prevented from applying their informal understanding. Pupils may create two independent systems of maths knowledge if mathematics training does not assist students in developing their formal knowledge from their informal knowledge. It is worth noting that pupils who get wrong responses on the written calculations can often get the proper answer by employing concrete materials. However, when challenged with the written word, around half of these pupils kept the wrong answer. Elementary students who had already connected decimal fraction number systems with physical representations of numeric amounts were more successful than pupils who had not built the same connections in resolving the problems they had never seen before, such as ordering numeric values by size and switching between decimal and common fraction types. Teachers give context to help students develop basic mathematics concepts and procedures, motivate them to discuss whether they are logical, and support them in making connections between primary and formal maths concepts and procedures.

2.7 Conceptual and Procedural Knowledge

Helping learners build links between conceptual and practical learning is one method to minimize such issues. Understanding the features of concepts, detecting the similarities among concepts based on these features, and creating the relationships between them are all necessary steps in forming thinking skills. Procedural memory, on the other hand, necessitates the development of skills, methods, or algorithms that serve as ways to reach a goal. For example, learners who may not match decimal points when adding or subtracting decimal fractions are likely to follow the procedure without connecting between decimal values and decimal point alignment.

More complex connections, such as adding up similar values, involve generalizing and reflecting on knowledge, such as aligning decimal points to combine decimal fractions or finding common numerators to add common fractions. Although such connections may be evident to grownups, children may find it challenging to make them. Teachers must provide lessons that assist pupils in constructing these large ideas. Student troubles can also be related to using improper representations in the classroom. For example, learners who are having trouble adding fractions may extrapolate incorrect algorithms from fraction representation training. Students frequently given fractions in the shape of pie graphs do "1/2 + 1/3 = 2/5" and defend their answer as "adding one piece of a two-piece pie to one piece of a three-piece pie might well result in two portions out of five total."

2.8 Developing Mathematical Knowledge for Teaching Elementary School Mathematics

While significant progress has been made in establishing a framework for students' mathematical skills, there is still more effort to be made in identifying what types of learning experiences are most effective in helping teachers obtain such knowledge. Mathematics teachers should first understand their students' present ideas in a way to construct them to assist pre-service teachers in developing math skills for instruction. The authors of The Mathematical Education of Teachers say, "The secret to turning even poorly equipped future elementary teachers into math learners is to work from what they understand" (Thanheiser et al. 2010).

Their attempts to construct math concepts from pre-service teachers' present beliefs are discovered to use two similar approaches.

1. For some, it helps determine pre-service educators' beliefs since they first enter school and then expand on those beliefs. Once the first concepts of pre-service teachers are recognized, and the growth of those ideas is understood, assignments can be developed to target those basic ideas, allowing for the development of more advanced systems.

2. Many help pre-service educators improve their concepts by restricting the mathematical notions that can be utilized in investigations; only concepts produced by the classroom are permitted. Each of these methods can be used at the same time. For instance, area equations for polygons might be established as a class based on the overall notions of surface and geometric attributes that pre-service educators have with themselves. The employment of either strategy (or both in combination) has encouraged the simultaneous formation of procedural and conceptual understanding while negotiating mathematical understandings in the classroom.

The studies emphasized involving pre-service educators in improving their understanding, facilitating chances for mathematical communication, and performing assessment activities to improve the knowledge to guide teaching while delivering subject classes for pre-service educators. This type of experience is essential for pre-service educators to ensure that they comprehend the teaching material and utilize it as a pattern for their education.

Finally, content knowledge is linked to and assisted by other categories of knowledge. Establishing linkages among knowledge domains is one of the goals of maths teachers' practice. Explaining teacher guides (linking to content knowledge and teaching), employing objects of children's mathematical thinking (connecting to subject knowledge and children), and discussing curricular decisions are all examples of teaching used by the writers to attain the purpose (connecting to the knowledge of content and curriculum). As a result of looking at our work collectively, we came up with a set of design principles:

1. Mathematical concepts are based on existing beliefs held by pre-service instructors.
2. Teaching comprehension should be modeled in pre-service teacher classes.
3. We concentrate on establishing links between subject knowledge and:
    - Knowledge of content and teaching.
    - Knowledge of content and children.
    - Knowledge of curriculum.

Such examples were chosen because they were related, highlighting four critical concepts in fundamental mathematics: place value, angles, the unit whole, and area.

2.9 Main sources of the Elementary Mathematics-Curriculum

The primary school maths curriculum, like other academic parts that make up the core curriculum as opposed to specialized or career path instruction, is characterized by three resources. The curriculum should be stable if and only if the three sources contribute equally, or at least evenly. The essence of the student, the nature of his or her everyday society, and the structure of the cognitive are the three origins of the primary school mathematics curriculum. The first is the psychological theory or the expressed needs of the child hypothesis of the curriculum (Callahan and Glennon 1975).

To a well-designed education, each has something to contribute. Each concept has strong and equal opponents within each group of individuals who are entirely uninformed that alternative points of view exist. Any solitary, dictatorial interpretation of the program's curricular foundation is radical. To have a comprehensive understanding of a balanced curriculum theory, one must have a complete sense of understanding of each of these radical theories. Each is discussed briefly here:

1. The psychological basis for curriculum theory: Curriculum theory's psychological origins and two psychological approaches to what mathematics is most valuable to primary school children can be used to answer it. The cognitive-developmental perspective is one, while the clinical personality perspective is the other. Either viewpoint is self-contained; one may rely on the other to varying degrees, depending on the viewer's professional training biases. The cognitive-developmental approach to curriculum theory focuses on the nature of the topic being understood and the learning process itself.

2. The sociological basis for curriculum theory: Those who promote a sociological approach to the selection of content for the elementary school mathematics program believe that the only mathematics that is worthwhile is that which has already been considered to be of considerable utility to the adult individual in business and general life situations. They claim that mathematical concepts that do not fit a strict interpretation of this condition are not a natural element of a child's general education. As a result, such themes form a part of the older child's or young adult's specialized or vocational education, to be learned in a vocational program either at school or on the job.

3. The logical or pure mathematical basis for curriculum theory: The logical, structural, or pure mathematical component is commonly regarded as the third source of the curriculum. Extremists who embrace this viewpoint are typically mathematicians who have little understanding of or concern for the viewpoints expressed by involving multiple psychological or sociological approaches. Their main objective is to communicate mathematics in an uncontaminated or undisturbed state, free of any attempts to relate the pure structure to socially practical problems. These three extreme places could be considered one of the triangle's vertices. A circle kept in place by springs, each held at a vertex, is seen as a balance among the three concepts. In this era alone, the demands of society on the education system have led the center of balancing to move frequently.

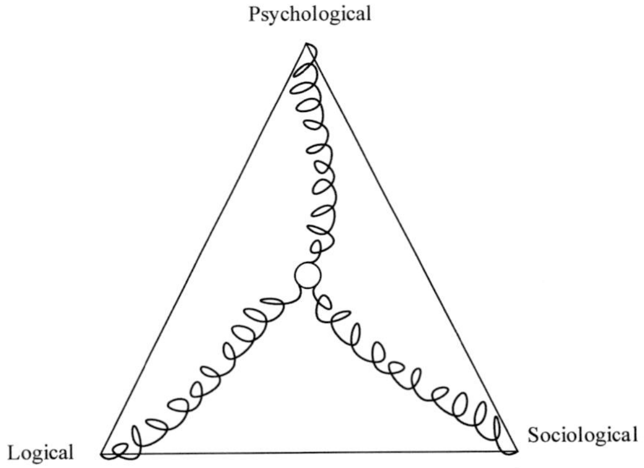

Figure 2.1 Models of Mathematics Curriculum Theories

2.10 Summary

Elementary school mathematics learning not only improves students' understanding and engagement but also improves their thinking skills. The chapter briefly described the five building blocks of learning mathematics for elementary students, along with the motivational and affective determinants in elementary school mathematics. It also discussed the student's difficulty in learning mathematics in elementary school and how to improve students' skills in maths by developing the content knowledge. The chapter elaborated on the case of elementary school mathematics and the primary sources of the elementary mathematics curriculum.

CHAPTER 3

INFLUENCE OF MATHEMATICS ON SECONDARY STUDENTS

3.1 Introduction

The learning and education of mathematics have a framework role that significantly affects schoolchildren's and pupils' knowledge acquisition and guarantees pupils' ability to use arithmetic in other subjects. During the 1980s, the secondary school mathematics curriculum saw a lot of changes, including a shift from descriptive teaching toward a wider variety of learning experiences, as well as a higher focus on problem-solving and experimental methods to assignments (Kyriacou 1992). Among the most recurring and prevalent issues in teaching, mathematics is that several pupils who excel in the field drop out as quickly as feasible, even though they are aware of the effects on their future employment. Researchers have tried to explain this phenomenon using various psychological perspectives, such as identity theory, self-efficacy, and leadership behaviors. Such research has been critical in changing the focus off from ability models, but it has generally not treated the phenomena as a cultural construct. The purpose of utilizing tests and other assessment measures during the educational process is to lead, manage, and evaluate pupils' understanding and performance toward achieving educational goals (Moyosore 2015).

Despite spending a significant amount of time in secondary school, pupils who do not learn mathematics are much less productive. Mathematics is an essential measure of a world-class education system that improves (Zakaria et al., 2013). Furthermore, it is an essential topic for developing countries because it allows students to pursue jobs in engineering, natural sciences, accounting, and many other fields critical to the national economy. Society views maths as the basis for technology and science, which is appreciated by individuals all over the globe (Miyazaki 2013). It serves as a political, economic, scientific, and technological progress mechanism. Mathematics is a topic that all students in secondary schools must study. Universities often utilize it to screen secondary school students for admission to elite science-based degree programs. As a result, this chapter examined how pupils might succeed in arithmetic and their transition from elementary to secondary school, among other topics.

3.2 Learning preferences and Mathematics Achievement of Secondary Students

Mathematics is a critical topic for promoting investment, especially in underdeveloped nations; nonetheless, learners score poorly in Mathematics compared to their peers. Considering the pupils' learning styles while educating is one technique to handle this situation. The level of achievement is a significant indicator of a country's educational system's performance. Furthermore, it is an essential topic for developing countries as it allows students to pursue jobs in engineering, natural sciences, accounting, and various other fields critical to industrial prosperity. The fact that scholars have stated that mathematics learning is in crisis is cause for serious concern.

3.2.1 Visual Teaching and Learning

Mathematics is a critical topic for promoting investment, especially in underdeveloped nations; nonetheless, learners score poorly in Mathematics compared to their peers. Considering the pupils' learning styles while educating is one technique to handle this situation. Student achievement is a significant indicator of a country's educational system's performance. Furthermore, it is an essential topic for developing countries as it allows students to pursue jobs in engineering, natural sciences, accounting, and various other fields critical to industrial prosperity. The fact that scholars have stated that mathematics learning is in crisis is cause for serious concern. According to the research, using such a vocabulary at school increased math knowledge.

Similarly, a teacher can aid visual learners' learning by utilizing proper Mathematics software that allows for dynamic visualization of topics. Given that studying styles vary and some students are multisensory, other learning styles must be acknowledged in the school. However, the way of learning appeared to be influenced by age and gender, with students in Sections 1 to 3 mainly relying on written than students in Sections 4 to 5, and girls being more adept at auditory, visual, and kinaesthetic learning than boys. This demonstrates that demographic factors can influence learning style preferences in mathematics.

Weak presentations and uninteresting educators who addressed the work in a dull tone and solely gave out the curriculum annoyed the students, mainly those who depended significantly on an auditory learning approach in class. Educators who did not check students' assignments resulted in no honest criticism and dissatisfied students who preferred reading and writing as active learning. Furthermore, some educators missed topic understanding and relied on their coworkers to help them clarify complex subjects in class. When questioned about which types of education in Mathematics classrooms work the best for students, the learners mainly mentioned reading/writing (as mentioned above), as well as hearing and group learning, Engaging on tests, and finishing assignments, exercises, and exams papers were all part of the reading and writing in the Mathematics lesson. Many also mentioned the need to interact with professors to resolve issues. Students admired teachers who provided thorough explanations and numerous examples, ensuring they grasped the material (Schulze and Bosman 2018).

3.2.2 Achievements in Learning

When asked how students learned Mathematics, the high performers' responses suggested that context influenced learning style preferences, in the view that distinct learning methods were utilized at home and school. Individual learning, as well as reading and writing, were frequently used at home, presumably given the fact that there was nobody to study with. A student who liked studying Maths with a friend made just one remark about group learning: Reading the notations in their exercise books and teaching materials, redoing theories that they had struggled with in class, researching other techniques of doing equations on the internet, working on exercises from their writings, exercising the instances and subjects from their syllabus, and having completed previous practice tests were some of the study strategies used by the students as they prepared for examinations. Only memorization was utilized to learn mathematical equations. When questioned about how the education of Maths could be upgraded, the students stated that teachers could develop the students' consciousness, ensure that the classrooms are comfortable to reduce anxiety, adequately explain, and be receptive to inquiries, etc., that would encourage auditory understanding. The students also proposed tutoring, a kind of group learning (Hossain and Tarmizi 2013).

Educators in the Mathematics classroom must be patient, courteous, and proactive in their guidance of the students. As the primary source of student conflict, the instructor is typically cited as the most common cause. Pupils who are anxious dislike providing ideas besides their friends, seeing such circumstances are potentially dangerous. Due to a lack of teaching time, teachers may choose to prepare their pupils for evaluation rather than learning. When students meet unusual difficulties, or the mathematics grows more sophisticated, this adds to the anxiety. According to the researchers, worried educators spend less time explaining maths and are more willing to pass on their anxiety to their students (Ng 2012). According to research, the teachers' remarks and critiques can considerably affect students' consciousness of Mathematics, which can cause their accomplishments. A review on the concern of remote secondary students in the State regarding Mathematics that limited their attitude, ambition, and progress also highlighted the need for a supportive atmosphere. As a result, the researcher advised teachers to establish "inviting academic situations." Mathematics teachers should be educated, encouraging, and patient, and they should help students develop positive self-concepts so that students believe they can learn independently at home. Many educators at the respective schools might lack the necessary subject knowledge to teach Mathematics adequately. According to research with Grade 10 students, improving teachers' content understanding improves learners' achievement dramatically. On the other hand, they help identify the association between learning style preferences and math skills in a sample of secondary school students and provide numerous suggestions for more successful learning and teaching.

3.3 The Construction of Identity in Secondary Mathematics Education

It attempts to figure out why sure pupils will maintain their advanced maths education while others would not. Researchers consider the concept of "identification" crucial to our investigation. According to the findings, pupils who create an identity that corresponds with mathematical discourse are more inclined to maintain school education than their colleagues who do not establish such an identity. Knowledge of the mechanisms through which learners construct a concept of who they are in connection to maths is crucial to the idea (Zevenbergen 2000).

3.3.1 Psychological Studies on Identity

Most identity formation research has been influenced by personal concepts, such as the notion of self-concept. According to these beliefs, there is a lack of knowledge of an individual's self-concept in connection to a sociological unit in the initial phases. As children grow, they become much more knowledgeable of who they are within such a society and, as a result, begin to investigate social norms. Individuals become more devoted and confident among their selected group when they become more conscious of the basis of their collective identity concerning specific other units (in terms of race, class, ethnicity, work, gender, and so on). These beliefs are founded on the maturity level concept, which states that as pupils grow older and mature, they will interact primarily with their classmates. Other theories take a more social psychology perspective. The method focuses more on a defined sense of belongingness, the related sense of security, and other behaviors linked with group identity. In part, belonging to a group is considered a fundamental feature of a building self-concept, where members develop a vivid awareness of the group's and organization membership's importance and, as a result, acquire great self-esteem through group membership.

Members with a favorable opinion of one's group have good self-esteem, whereas those who have a negative perception of the group have poor self-esteem, according to a study on the effects of individuality in early childhood, and much other research has suggested there is a connection among self-image with a team and self-esteem. Such theories have highlighted three components in seeking to describe and assess "identity":

- a feeling of belongingness
- a feeling of fulfillment within the group's standards
- behaviors linked with affiliation to a particular group that is considered significant features of self-image

Such elements are markers of essential characteristics to examine while thinking about personality from the psychological perspective. However, to fully understand that these characteristics are displayed, it is necessary to study the social situations in which they are formed.

3.3.2 Mathematics as a Community of Practice

A cultural perspective on identity, in contradiction to psychological views of individuality, considers how learners engage with their social world and how the two characteristics, the individuals and the organization, are jointly significant to individuality. Humans develop a sense of identity and purpose only via social practices and connections rather than understanding as a process that happens "inside" a person. People will not have one personality but multiple identities prominent in various contexts, according to the repositioning of individuality as a consequence of engagement in different groups. As a result, identification is portrayed as changeable and placed rather than stable, constant, or life-long. Researchers have methodically investigated the interaction of society, action, purpose, and personality by their research. For researchers, it appears to become a more constructive way of understanding not only how learners know mathematics but also how they understand math concepts and gain a sense of whom they are as learners inside the social process of mathematics by getting involved in a community of inquiry, in this particular instance.

3.4 Transition from Primary to Secondary School Mathematics

Several pupils undergo substantial effects on the physical framework, teaching and learning procedures, and educational objectives when they move from primary to secondary school. Transitions to high school occur when pupils are between the ages of 11 and 12 and go through the physiological, mental, and social changes that come with youth. According to research, challenging transitions have been linked to disengagement, unfavorable emotions about the school, low self-confidence, and low motivation, notably in the domain of mathematics education. Detachment with mathematics can limit students' ability to interpret life events from a mathematical perspective and reduce the number of higher education courses obtainable to them (Attard 2010).

3.4.1 Middle Years, Mathematics and Transition

Within and inside the school, as well as outside the mathematics teaching, are the factors that can affect students' mathematics involvement. Although the move to high school can have a significant impact on involvement, many other aspects are unique to the teaching and learning of equally important maths. Education, methodology, evaluation processes, human interactions, and pupils' interactions with others are such elements. When combined with transition, the occasionally harmful effects of these elements are reasons for concern. Over the last two decades, studies have consistently shown that a lower percentage of pupils study mathematics at the higher secondary level and beyond. Students' perspectives toward and achievement in mathematics, which are affected by their school mathematical activities and the training they received, have significantly impacted their decision not to continue in mathematics. Negative views toward maths are challenging to alter once developed and can remain into adulthood, even though views may shift throughout the educational career. Maintaining interest in math during the later years might develop more positive views, making maths education more appealing.

As they transfer to high school, pupils undergo social, cultural, and educational adjustments. Students ready to move on from elementary school frequently have preconceived ideas and unrealistic expectations about the obstacles that secondary schools will provide. Several Year-6 pupils assume that the work in Year-7 will be more difficult, creating new challenges for some while causing anxiety and concern for someone else. Students thought academic studies in their first year of secondary school were not significantly more difficult than in their final elementary year, but they struggled to adapt to the new educational setting. Despite the lack of competition, the move to secondary school frequently results in a loss of performance. Students encounter considerable social adjustments as they transfer to high school regarding the academic problems mentioned above. Many pupils must find a way to deal in a much wider school setting, with a strong focus on control, more professional student/teacher connections, and a higher possibility of public assessments of students than in primary schools.

3.5 Active Learning in Secondary School Mathematics

3.5.1 The Nature of Active Learning

Practical work, machine learning, roleplaying game exercises, job experience, customized work plans, group debate, interactive problem-solving, and extended work are all examples of active learning. Think about such a list and ask, "What can they have in general that could explain active learning?" Learning by doing, learning by experience, learning through action, learning through talk, student-centered learning, peer cooperation, and cooperative education are some of the terms used to describe active learning (Kyriacou 1992). Some organizations have tried pinpointing the essential elements or qualities underpinning active learning. Two key qualities, for example, are a concentration on studying by doing and a focus on student choice. Students should be permitted to ask their particular inquiries and use educators and other facilities to achieve self-defined purposes through active learning. The seven fundamental principles of active learning are:

- reflective: the student begins to question the significance of what is being learned;

- purposeful: the activity is regarded by the student as related to his or her issues;
- essential: the student appreciates many frameworks for understanding learning;
- negotiated: the instructor and the student negotiate the academic objectives and techniques;
- challenging: the tasks are challenging in nature;
- scenario-driven: learning activities are based on the needs of the circumstance;
- engaged: learning processes are based on real-world tasks.

The first four components are categorized as participation features, whereas the remaining three are categorized as realism factors.

Active learning is defined as the use of having to learn the process in which learners have a high level of ownership over the learning process used, the learning opportunity is open-ended rather than tightly predetermined, and the student can participate fully and contour the educational experiences.

3.5.2 Active Learning in Mathematics

Analyses of active learning in mathematics present several criteria from relatively straightforward studies examining active learning as essentially designating specific instructional practices such as group activities and computer-aided learning. These studies frequently aim to discover the characteristics of good practice that help achieve the desired educational outcomes. However, some research highlights more fundamental problems regarding mathematics, characterizing mathematical activity, and what defines mathematical learning. Many authors situate their views on active learning inside a framework focused on the structure of the mental activity that is taking place, which is most typically a constructive version of mathematical learning. Educators are said to encourage higher student engagement and enthusiasm, encourage students to explain math concepts, boost confidence, and provide more effective learning.

Educators in elementary and secondary schools are asked to choose from various potential classroom activities that occurred during their last session in a randomized class. Many high school mathematics sessions are still mainly pedagogical, focusing on presentation, discourse, and assignments provided from the handbook, according to the results (occurring in 89% of lessons). Nonetheless, there was confirmation of a substantial amount of work done in small groups (40%) and, to a smaller extent, the use of hands-on or interactive resources (16%), as well as the use of pcs (8%). Moreover, the perceived effectiveness of small group work may represent nothing beyond students cooperating, while active learning necessitates much more.

3.6 Self-Concept, Self-Efficacy, and Performance in Mathematics

Consciousness has been identified as an essential self-perception in a wide variety of educational circumstances, according to research. Consciousness refers to the belief in one's ability to plan and carry out the steps necessary to attain specific performance goals. Beliefs are widely used in academic settings as a considerable influence on academic achievement and as a moderating mechanism impacting pupils' levels of effort, tenacity, and endurance, as well as their mental responses and self-regulation. According to research in various academic contexts, improving efficacy beliefs is essential for achieving practical goals, such as greater levels of learning in writing, maths, and general educational achievement (Pietsch, Walker, and Chapman 2003).

3.6.1 Affective Component of Self-Concept

The relative importance of each concept is one variation. Self-efficacy beliefs are predominantly intellectual judgments of skill, whereas self-concept beliefs, as tested by the Self-Description Survey, incorporate psychological and emotional evaluations of mastery. Although there are a variety of conceptual theories of self-concept, several of which split the cognitive and emotive elements into two independent structures, researchers employing a variety of self-concept measures have included personal items in their self-concept assessments. Compared to domain-specific self-concept, efficacy beliefs connected to diverse performing domains deal mainly with cognitive views of competence and are developed by thinking about enhancing learning activities, vicarious learning, precursory information, and physiological evidence.

Subject-specific self-concept views may be distinguished from practical ideas tested at the same degree of abstraction by the presence of a subjective element within beliefs. However, the concepts of being that contribute to the formation of subject-specific self-concept beliefs are expected to be intertwined with performance judgments. Within learning settings such as mathematics, concerns (efficacy beliefs) and problems (self-concept beliefs) may have significant theoretical overlap. As a result, when subjective components are eliminated from self-concept assessments, the resulting self-concept items and self-efficacy beliefs items are more likely to concentrate on a single factor. Many types of research have shown that self-concept, as examined by the Strengths and Difficulties Questionnaire (SDQ), has two separate components, with supportive evidence for the presence of competence and effective response within each subject-specific self-concept.

3.6.2 The Role of Social Comparison Information

The purpose of social evaluation in establishing self-concept and efficacy beliefs is the subject of a second conceptual contrast. The self-concept is shaped in part by other people's reflecting evaluations. On the other hand, self-efficacy is dependent on mastery beliefs rather than prescriptive requirements. Although subjective knowledge has been identified as a form of adequate data, prior mastering knowledge is the most crucial source of adequate data. Despite that, efficacy theory emphasizes the importance of comparison processes, such as experiences, in constructing efficacy beliefs.

On the other hand, self-efficacy researchers claim that efficacy beliefs emerge from a different type of social knowledge. Efficacy beliefs are formed through vicarious experience, which involves analyzing individuals' performances in developing one's ideas of skill. The construction of effectiveness beliefs is aided by cognitive assessments of one's cognitive likeness to the individual performing the activity and the relative difficulty of the activity completed.

The comparative significance of social evaluation in constructing the two conceptions may depend on the amount of specificity at which they are evaluated. In situations where previous experience is not widely obtainable, general judgments of ability can rely heavily on comparison processes than on previous experience. Pupils in mathematics, for example, have no previous experience with the maths they are currently taking. Learners cannot rely on past mastering experiences in these specific courses to predict the self-efficacy for having to complete their mathematics course. However, pupils can develop their estimates of ability at higher levels of generality by using relevant knowledge about their value ratio in their class or year at school. As a result, pupils' opinions of their relative strength with their classmates may have an equal effect on global efficacy for maths and mathematics self-concept.

3.6.3 The Predictive Utility of Self-Efficacy and Self-Concept

The facts regarding self-efficacy and the self-predictive concept's value reveal that the causal association connecting self-efficacy and educational success is more reliable than that of self-concept. Self-efficacy and self-concept must be quantified at generality levels equivalent to the achievement indicator's granularity level. However, research including self-concept and self-efficacy assessments has generated conflicting results. Much investigation into self-efficacy and self-concept has evaluated self-efficacy and self-concept utilizing different measures developed independently for every structure, assessing both concepts at various levels of generality to equate the predictive value of self-efficacy and self-concept.

The component is assessed at the same granularity level as the performance usually has a higher predictive ability. Furthermore, similar research has included an evaluation or subjective element in their self-concept assessments that would be less important for academic success than the competency element. However, as stated previously, a theoretical difference is created between self-efficacy and competence-related self-concept ideas. According to the researchers, efficacy views are established by addressing questions, but self-concept attitudes are developed by questioning. However, there seems to be significant paradigm overlapping between conceptions; it needs to be seen whether structuring skills in activities that can be completed or individual capacity is more clearly connected with achievement goals. Since efficacy assessments need a response referring to self-confidence in reaching specific outcomes instead of evaluating individual traits about general ability, self-efficacy beliefs must be more strongly linked to academic accomplishment than self-concept.

3.7 Perceptions of Mathematics Education Quality on Mathematics Achievement

The concept of educational quality is identical to the basic concept of quality. The total of the aspects that determine the service or product, based on the satisfaction with the defined needs, is known as quality. As a result, the first group of researchers who employed this notion defined performance as educational excellence. Quality education refers to how training is structured, handled, and delivered, what is delivered, the level of knowledge, and the particular instructional objectives accomplished. The notion of educational excellence has sparked the implementation of influential organizations and set the stage for various studies. A successful school is one in which anyone in the school environment has a clear goal and is committed to teaching purposes, prioritization, evaluation techniques, and responsibility. Good leadership, human resources, a supportive environment, community, and pupils who are driven to learn all contribute to the success of a school. Excellent schools have strict requirements and prominent, approachable, and fair participants (Ciftci 2015).

3.7.1 Education quality of Mathematics and Student Attainment

According to researchers, it includes using several kinds of assessment to guide and improve a student's learning performance. Informative assessments, a pupil takes diagnostic tests at different phases throughout their studies and receives counsel (remediation) on how to continue next based on the findings of the tests. Pupils' behavior is likely to increase if their knowledge is effectively assessed daily during the teaching-learning phase and fast evaluation and intervention are offered. Student performance is one of the indicators of educational excellence Mathematics was revealed as the most challenging area whenever pupil results were assessed in terms of curriculum. Mathematics, which began in elementary school, has become a nightmare for learners at all levels of study. Trends in International Mathematics and Science Study (TIMSS) and Program for International Student Assessment (PISA) are two international educational assessment research that provides information (James and Folorunso 2012).

The ability of learners to experience satisfaction and be protected in their schools, as well as favorable opinions of the training they receive, is among the most critical components of their educational, sociological, and mental development. Schools, the key accountable organizations in comprehensive education, are intertwined with various elements to fulfill such tasks. For example, all ideas comprise students' achievement, which may be read and understood by investigations such as TIMSS and PISA, which represent student views towards the curriculum; therefore, student perspectives and uncertainty regarding these classes are echoed in their academic achievement. Today's learners expect more from education than just academic success; the values of school career and other activities (such as moral, social, athletic, commuting, and technological) all contribute to the growth of a good attitude regarding school among students.

The preliminary test, in particular, allows pupils to compare their achievements to others. It enables for easier handling of pupils, educators, and academic achievement simultaneously. It is valid to compare the performance of a class or an institution to the national median both on a student level and a class or school level. Parents are urged to analyze the factors that contribute to their children's success and the spaces used at the educators' demand. School administrators can evaluate teachers' performance, while the administration can evaluate academic achievement, resulting in more efficiently allocating of resources. In terms of educational monitoring, a positive association is thus created between national exams and student achievement.

3.8 Summary

Due to the importance that mathematics assumes, the subject became central to the school curriculum. The mathematics curriculum is intended to provide students with essential knowledge and skills in a changing world. Mathematical ability is also crucial for the economic success of societies. Mathematics skills are essential in understanding other disciplines, including engineering, sciences, social sciences, and even the arts. Thus this chapter elaborated on the construction of identity and how students prefer their learning style in secondary mathematics education. It also discusses the student's experience in transitioning from primary to secondary school mathematics. Also, the relationship between self-concept, self-efficacy, and performance in mathematics during secondary school is briefly explained.

CHAPTER 4

OUTLOOKS OF SENIOR SECONDARY STUDENTS ON MATHEMATICS LEARNING

4.1 Introduction

Mathematics is an essential skill in society, especially in this technological age. It is no surprise that Maths is compulsory in secondary schools, even though not all pupils are required to become mathematicians. It is due to its widespread use in daily life. Functioning well in one's surrounding environment requires mastering mathematics teaching. Mathematics is regarded as a fundamental factor in science learning, such as technical and industrial, as well as good analytical thinking in everyday life in modern civilization (Oluwatoyin 2007). Science has been regarded as a medium for technological, cultural, and economic progress across the planet. Mathematics is not only fundamental to such disciplines but also the application of science. Mathematics is a topic that pupils in higher secondary schools must study at least four times one week, given its importance. It is done to ensure that pupils thoroughly understand the topic.

Mathematics is for life, and everyone uses it in some form or another in their daily lives. Mathematical knowledge is more important than ever before, particularly with the modern problem of science and technology growth (Jebson 2012). The importance of maths in the education of higher secondary schools could not be overstated because there is no clear, all-encompassing definition of maths. Because of its generality and diversity, mathematics signifies several different things to various individuals. Calculation, algebra, and trigonometry are mathematics divisions described as the knowledge of quantities and numbers. According to the encyclopedia of maths, mathematical knowledge of numerals and spatial shapes in the real world is intertwined with technological and natural scientific needs. As a result, maths is a science that compacts with the symbolism of concepts and relationships. It can also be viewed as a tool for conducting a conceptual investigation to put diverse concepts into practice. Mathematics is a study that is required at all stages of modern schooling institutions and is helpful in everyday life.

4.2 Increasing Senior Secondary School Mathematics Achievement

Mathematics teachers generally strive for learners' homogeneity regarding talents, creativity, skills, and ability to generate equal academic achievement. Although that desire, massive differences in understanding, talents, attitudes, behaviors, cognitive abilities, thinking mechanism, passions, learning speed, learning preferences, linguistic competence, academic motivation, and progress emerge among students in mathematics classes. In mathematics courses, despite these independent sectors, learners are expected to take part in the acquisition of ideas, methods, and abilities under the supervision of educators (O. A. Awofala and O. Lawani, 2020). Learning might decrease when a teacher presents his or her lessons in a class without considering the unique peculiarities of the pupils' mathematics. It is common in mathematics classes, where educators are obsessed with delivering homogenous facts and knowledge. Teachers could achieve equality in the school setting by finding a strategy that gets into their pupils' variety, which can be accomplished using varied instruction.

4.2.1 Differentiated Instruction

Differentiated instruction is an educational philosophy that holds the instructional approaches that should change and be altered as per the needs of different and various learners in the class. Differentiation is a method of instruction in which educators predict and adapt their instruction to meet the specific requirements of potential pupils and small groups of learners to maximize each learner's academic potential in the school by employing rational methods for educational quality observation and information outcome. Differentiated teaching requires teachers to explore a range of student characteristics while planning and creating sessions and divisions, demonstrating that it is not a single method but a system of education that incorporates a variety of methods. Differentiated instruction requires educators to be flexible and responsive to lectures and tutorials, whereby the unique student aspects in mathematics classrooms are considered when preparing and providing teaching methods.

Differentiated instruction is a system for adjusting instructional tactics to enhance each pupil's knowledge preparation, areas of passion, and learning patterns. It consists of five key elements: the topic, the procedure, the result, the impact, and the educational experience. The learning, ideas, abilities, behaviors, attitudes, generalizations, principles, and behaviors pupils must learn in class are content. Teachers must concentrate on the ideas, methods, and abilities that learners acquire since content refers to what is explained and how it is conveyed to pupils. The techniques that students use to get meaningful content are what educators can vary in terms of content. All students need to be taught the same principles, but the level of complexity needs to be adjusted to suit different types of pupils. Teachers are focused on creating knowledge activities to help learners take charge of the subject by allowing students to understand how the subject makes sense and identify how the content is appropriate and relevant outside of the school due to the differentiation process (Fatade et al. 2014). The process is a basic level in differentiated education since, during this phase, individuals learn at different paces, with varied types of support, in various groups, and with various learning methods. The teacher gives differentiated instruction, whereby all pupils learn together at their own pace, with below-average pupils using secondary tools, and the teacher offers an additional challenge to students based on their performance throughout the class.

4.2.2 Achievements in Differentiated Instruction

Differentiated instruction has been proven more efficient in encouraging and promoting pupils' performance in mathematics than traditional teaching methods. It allows pupils to study together in flexible groups and trains them to acknowledge the characteristics that could inhibit their mathematical growth. Differentiated education considered several differences in readiness levels, preferences, and learning styles. The teacher provides the pupils with a way to strengthen abilities and ideas using maths subjects they enjoy learning by considering their interests. Teachers may compensate for the pupils' seeing and listening, thinking and behaving, logical and instinctive thinking, and evaluating and visualizing inclinations by looking into individual preparation levels. Teachers could utilize mathematics curricula in innovative lessons by identifying and arranging mathematical thinking based on learning purposes, choosing educational methods for the appropriate transaction, and conceiving formative assessments based on learners' interests, study habits, and preparation levels.

Educators in the differentiated instruction group may have achieved higher performance levels because they were subjected to new experiences that engaged them in an interactive way of recognizing connections among ideas in which new information was resolved, successively differentiated, and well-incorporated into previous knowledge. Similar research has linked differentiated instruction's efficiency in enabling significant concept knowledge and improving pupils' mathematics performance. Students demonstrate knowledge and control over the topic when ideas are effectively learned and absorbed, resulting in considerable improvements in student performance. However, several researchers have found that differentiation does not influence pupils' academic performance in school concepts. The underperformance of the group of students, as evidenced, can be related to the flawed nature of the traditional teaching style, in which the pupils are just passive learners in the learning experience, depriving them of the ability to participate in the learning process.

4.3 Problems and Interventions in Senior High School Mathematics

The "Enhanced Basic Education Act of 2013," also known as the K to 12 Curriculum, changed the educational landscape from a 10-year primary education to a 13-year required education that included Preschool and two years of Senior High School. University preparation, vocational schools career aspirations, visual arts, sports, and entrepreneurship employment, are all aims of a high school degree. Intellectual, Technical and vocational Livelihood, Arts & Design, and Athletics are the four disciplines of Senior High School (SHS). Accountancy & Business Management (ABM), Science, Technology, Engineering & Mathematics (STEM), Humanities & Social Sciences (HUMSS), and General Academic Strand (GAS) are the four strands within the Academic Track. Nevertheless of their chosen career path or educational stream, all SHS students need to take both core or basic school mathematics subjects: General Mathematics and Statistics & Probability. Furthermore, specialized Maths classes are specific to the career path or education path chosen by the students, such as Business Mathematics for ABM and discrete mathematics and essential calculus for STEM (Jaudinez 2019).

The Maths Curriculum Guide emphasizes the importance of students studying and exploring maths clearly and in detail since its significance extends further than the classroom. As a result, an Educator needs to engage, enable, and motivate learners to attain maths goals (critical thinking and problem-solving), mainly in Senior High School, the concluding stage of primary education. Senior High School is designed to allow students to strengthen their educational skills to prepare them for their careers better. The STEM program was poorly received by major educational participants such as pupils, parents, educators, and administrators. The particular academic thread is likewise seen to be tough to work with. The lack of facilities and teaching aids was still a severe issue in carrying out the curriculum's objective. Students' pre-requisite abilities were also determined in preparations for Senior High School Maths.

4.3.1 Teaching General Mathematics

When the lesson begins, the educator connects some principles to the students' everyday activities, such as fares from a "function" of distance covered. It is also easier to teach the first volume, which is about functions, as pupils have been subjected to it since eighth grade. The use of boards and seating works as a fundamental evaluation tool to monitor performance is still in use. Learners are provided with roughly 4 to 6 exercises for understanding and knowledge, depending on the complexity of the content of the lesson. Moreover, because the performing assignment in the instructional guide takes a long time, combining points/ratings from the lecture notebook, attendance, and portfolio is an alternative solution for evaluating students' success. Because educating Mathematics typically requires a longer duration of discussing a problem's solution, pupils have almost no chance to clarify or ask doubts and justify critical properties.

4.3.2 Teaching Probability and Statistics

The educator first goes over essential ideas in a specific lesson, and formative assessment is typically done through oral questions and lectures. Once the subject has been clarified, a question may be asked, or students may be encouraged to participate in a group activity or do a performance-based assignment. For example, pupils may be requested to survey the house or class and then announce the findings in class. Students used MS Excel to compute statistical analysis and visualize data graphically. Making or drawing diagrams on paper is an option for those without PCs. When presenting, all students have the chance to speak about the results with the teacher's monitoring. Calculators also prove to be helpful when it comes to calculating descriptive statistics. It has been noted that when pupils deliver their work in their language rather than English, they do so very well. The activity's numerous topics have been maximized due to puzzle grouping. Statistics and probability are taught engagingly and experientially.

4.3.3 Teaching Precalculus and Calculus

The lecture is first written on the board, then copied into the pupils' notebooks. The teacher then explains topics and methods to the students with only a few oral questions. The teacher frequently demonstrates a methodical approach to solving questions in calculus, including limits, continuity, derivatives, and numerical methods, with pupils seated and going to understand the teacher's presentation of the topic. A quick quiz is usually given after a class. The educator encourages learners to complete the exercise at home to develop analytical ability. The need for memorizing formulae was also underlined. Conic segments and charts of algebra and transcendental equations are also thoroughly depicted to give an exact visual image of the curves. Conic segments and charts of algebra and logical function were also adequately represented to give an accurate visual image of the curve. These would have to be drawn on blotting papers by the students.

The lack of understanding of critical skills, remaining stigma, and language utilized have all been faulted for students' underperformance in subjects and learning, computational thinking, visualization, problem-solving, and other math skills and abilities. It could be worsened if educators continue to use unfriendly and inaccurate books and other materials.

4.4 The Reasons for Poor Math Performance in Senior Secondary School Students

Various factors can cause poor mathematics ability in higher secondary school pupils. Some categorize the factors that work against substantial academic success into four main categories:

- Causes in the student's environment include basic cognition skills, physical and health problems, psycho-emotional factors, and a loss of motivation in school.
- Family factors include cognitive stimulus or fundamental perception in the first two years, home regulation, absence of a leadership role, and finances.
- Causes include the school's environment, physical structures, and interpersonal relationships among the staff.
- Destabilization in educational policy, underinvestment in the educational system, leadership, and career losses are all examples of societal causes.

Numerous researchers and authorities have identified various reasons for pupils' low mathematics achievement (Sa'ad, Adamu, and Sadiq 2014). For example, there seems to be a belief that a lack of well-trained educators, insufficient teaching facilities, insufficient funds to buy equipment needed, poor handbook quality, class sizes, low motivation educators, a lack of labs and library resources, poorly organized supervisory activities, public service meddling in the education system, unceasing exchanges of educators and administrators, involuntary promotions of students, and the significant role of public assessments are all factors.

The noticeable causes of poor performance in mathematics are:

- There is a severe shortage of professionally trained math teachers.
- Most mathematics professors have demonstrated a lack of understanding of teaching mathematics.
- Overcrowded math classes.
- Bad attitude of pupils toward maths.
- An excess of attention on curriculum completion at the risk of genuine understanding of maths ideas.
- Short maths labs and supplies.

The reasons for poor performance in mathematics from the point of view of principals are:

- Lack of education support
- Head teachers' disappointment with in-service mathematics teacher training
- Perceived shortage of instructional resources for teaching mathematics
- Pupils are being educated by teachers who have not finished any career development during their professions.
- The maths curriculum was not entirely covered. A particular emphasis is focused on a few numerical areas.

Some factors that contribute to low mathematics achievement among higher secondary school pupils provide an impression of mathematics as a complicated subject and anxiousness. In schools, students typically develop mathematics stress studying from educators concerned about their mathematical skills in specific areas (Umameh 2011). Parental attitudes, inconsistent coaching, bad teaching, and cognitive deficits contribute to poor maths achievement. The lack of valuable libraries and laboratories, competent educators, home environmental conditions, family circumstances, and inadequate parental involvement in their children's education are the primary reasons for poor math achievement. As a result, it is evident that the causes of low mathematics achievement in higher secondary school students are numerous and varied, but it varies into three categories: education causes, teacher-related causes, and student-related causes.

4.5 Approaches for Improving Mathematics Performance in Senior Secondary School Students

Teachers, pupils, and organizations all have suggestions for enhancing pupils' mathematical skills. Every area of human thought recognizes the relevance of maths in research and technology. Hiring, teaching, learning resources, syllabus, motivations, and taxes and assessments can help pupils improve their mathematics achievement. The formation of a good attitude toward maths, the management of more assessments and questionnaires, the requirement of teaching and learning substances, encouragement, critical phases of the curriculum, the availability of adequately trained teachers, use of a variety of educational, and the ability to monitor of lessons by the school staff are all ways to improve pupil achievement in mathematics. The requirement of teaching materials, library, laboratory, and other facilities, as well as the development of good bonding with parents by school principals and the reduction of the student-to-teacher percentage to a manageable number, are some ways to enhance mathematics achievement.

Pupils' mathematical performance can be improved by the use of the following four approaches: when teaching mathematics, groups into students' ability should be imparted; the constructionist technique might have been imbibed in mathematics education, that is, for learners to study and maintain their knowledge, they must be in charge of their studying. Also, it uses instructional aids and activities, as well as computer-aided teaching. Some more ways may be employed to help students boost their math ability.

- Reframing pupils' perceptions of mathematics as a challenging subject.
- Reducing the detrimental statistics of a fail rate in mathematics tests.
- Increasing awareness that maths is at the root of technological advancements.
- Assuring pupils pass mathematics to enter numerate academic subjects for further study.
- Establishing a genuine interest in mathematics in children as early as kindergarten.

- Establishing a trustworthy platform for recognizing, supporting, and awarding excellence.

As a result, numerous approaches can be used to strengthen higher secondary school students' mathematics achievement, including the availability of good teaching and facilities, skilled teaching staff, the development of a positive mindset among students toward maths, parental support in their children's education, the use of proper teaching methods, the organization of quiz questions and competition between students and so on.

4.6 Perceptions of Difficult Concepts in Senior Secondary School Mathematics Curriculum

Some concepts might be challenging for those with poor logical thinking and analytical skills. These pupils would have visual or dyslexia difficulties, preventing them from perceiving patterns. On the other hand, students with excellent reasoning skills and knowledge may be competent in understanding some mathematical ideas but may find others difficult. The term "difficulty of concept" thus refers not only to a student's incapacity to pass a set of mathematics problems but also to what defines a "chronic hitch" that makes basic processing of a mathematical notion an arduous effort. Pupils often have fundamental trouble in math skills, mathematical concepts, and learning basic mathematical principles when identifying their issues with mathematics concepts. According to the experts, the majority of the trouble in interpreting symbols is that symbols can have different meanings in different contexts. In maths, knowledge necessitates following an algorithm that takes the solvers through a proper process to a valid answer (Charles-Ogan and George 2015).

For a framework and efficient process, pupils should be permitted to actively engage in each stage of a problem-solving procedure during education. Improper representations and treatment of problems, such as decimals, ratios, projection, and incorrect algorithms, might be faulted for some pupils' difficulty. Researchers have noted some difficulties among pupils, such as academic stress and attitude towards learning maths, as inherent in students (Mahmood 2011). The concept of an attitude refers to a person's way of thinking, acting, and behaving and significantly impacts the student. Educators who have a good attitude about maths, on the other hand, might encourage their pupils to have a good attitude towards maths. The pupil's attitude toward an educational process, whether natural or learned, affects their classroom behavior and mental attitudes towards maths. From various perspectives, a person's attitude toward mathematics was considered a more complicated phenomenon described by the thoughts associated with maths, personal views about math, and how someone acts toward mathematics.

4.7 Self and Cooperative-Instructional strategies for Improving Senior Secondary School Students' Attitudes toward Mathematics

Attitude is a situational willingness to pay attention to unavoidable circumstances, people, or things learned and become one's regular action pattern. The total of a man's preferences, emotions, discrimination, prior beliefs, thoughts, concerns, challenges, and convictions regarding any topic is referred to as attitude. Thus, an attitude is a view toward a person, item, activity, or idea that might be good, bad, unfriendly, or neutral. One's mentality can influence their actions and results. The formation of a good attitude has captivated people's interest since the manner with which one conducts any task or attempt has much to do with the results anyone achieves (Kundu and Ghose 2016). A positive attitude assists education, but a negative attitude hampers studying. Any Mathematics curriculum should aim to develop a positive opinion about maths and transmit cognitive information. However, higher secondary school students' widespread belief and frequent phrase that only brilliant pupils can effectively master Mathematics leaves something to be desired by senior secondary school Mathematics curriculum developers. These beliefs have influenced pupils' attitudes toward Mathematics and their academic ability in the subject.

Higher secondary school students frequently have a poor opinion of mathematics, which leads to a lack of motivation in the topic and, as a response, reduced academic performance in Higher Secondary School Mathematics Examinations. In conclusion, math education scholars have emphasized and suggested using various instructional strategies to promote a positive approach to the field. Music, project-based learning, clubs and activities, self-regulatory learning, and advanced organizers are just a few excellent strategies that have improved students' attitudes toward mathematics learning (Ifamuyiwa and Akinsola 2008).

4.7.1 Self-Instructional Strategy

Individual learners utilize educational packages or learning packages (mainly designed educational packages that include inputs, availability for answers, information, and assessment) with which learners can study either without instructional support or with a minimal educator. Self-instruction concentrates on each learner's particular requirements, talents, learning styles, interests, and academic background and provides opportunities for them to improve and progress. Teaching strategy is a teaching method that allows most of the learners in a class to work at their own pace while receiving answers quickly. The central assumption is that when students are in charge of instruction, they have more time to think thoroughly, arrange the learning activity, and decide on the significance of what they have learned.

The resources are organized and presented in a learning package that leads learners from a collection of recognized fundamentals to the unfamiliar, from basic to complicated concepts within the same academic sector. The effectiveness of self-instructional strategies in increasing the cognitive and emotional educational objectives of students in various fields has been documented in numerous studies. Despite the current conceptual emphasis on learner-centered types of instruction, educational practitioners have not accepted self-learning' as a suitable teaching design for their elementary and secondary schools. The current study, which used a self-learning package, is an effort to investigate the efficiency of personalized mathematics instruction further, particularly when contrasted with cooperative learning strategies.

4.7.2 Cooperative Learning

One of the most fundamental assumptions about how students learn maths is that they learn alone or together. Active learning during the teaching-learning process must allow pupils to interact with others, whether they realize their maximum potential in maths, particularly at the higher secondary school level. The cooperative-instructional technique is one learning model that quickly provides students with a collaborative learning environment. Such a technique is wholly opposed to the self-learning method, which emphasizes students' ability to learn individually. Cooperative learning is a pedagogical practice that promotes or motivates students to collaborate on academic assignments. Cooperative-instructional strategies can be as basic as allowing learners to sit together in a circle to talk about or assist others with classroom duties, or they can be pretty complicated.

Teachers agree that using a cooperative-instructional technique improves learning results. Researchers discovered that in classes where students work in pairs and larger groups, mentoring and sharing prizes, students have a more robust comprehension of information than in schools where students recite material individually.

There is mutual support among pupils' goal accomplishments in a cooperative learning environment; pupils believe they can achieve their educational objectives if and only if fellow pupils in the learning group equally achieve their goals. The cooperative learning hypothesis is founded on the idea that when people work together to achieve a shared goal, they rely on one another's efforts to succeed. Individuals, like many, lead to positive outcomes, and cooperation usually increases positive contact between many group members, so cooperation in a cooperative approach encourages teaching and learning to inspire one another to do whatever tends to help the group succeed. One of the most beneficial aspects of cooperative learning is its social training. Pupils understand how to cope with disagreements, appreciate those with differing viewpoints, collaborate to accomplish a more significant result, and function as a team through working in a group. Learners know the sense of community that emerges from becoming a part of a group and how to handle the joy of success in challenging assignments and the aches of failure maturely through teamwork.

4.8 Summary

Mathematics improves pupils' problem-solving and analytical capabilities and enhances their logical, functional, and aesthetic abilities. In general, humans use mathematics daily, and this everyday use of mathematics causes the human brain to express issues, ideas, and solutions necessary for the human race's existence. Thus, this chapter enlightened me about increasing Senior Secondary School Mathematics Achievement and how to teach them maths using problems and interventions. It also deliberated on the cause of their poor performance, ways to improve their knowledge in maths, and self and cooperative-instructional strategies for improving senior secondary school students' attitudes toward mathematics.

CHAPTER 5

CURRICULUM AND PEDAGOGY IN MATHEMATICS

5.1 Introduction

The early childhood mathematics curriculum has been the subject of much current research and innovation. For example, producers have lately generated a wide array of novel preschool programs, re-stimulated by research revealing that performance disparities among students from low and more extraordinary regions start in the early years. Several individuals may be led to conclude that childhood development; math is a novel phenomenon due to these bursts of activity. Tradition, on the other hand, demonstrates that math and disagreements over the types of mathematical skills that should be taught have a long tradition in prior learning (Cobb and Hodge, 2002).

Diverse perspectives on the suitability of math for young kids led to disagreements. Critical viewpoints were frequently based on vast social ideas or trends rather than research or observation of children. Traditionally, those who dealt with young kids created a valuable curriculum framework. Mathematics, for instance, was present throughout Friedrich Froebel's work as the pioneer of kindergarten (which initially comprised students from 3 to 7 years of age). Froebel's primary abilities are primarily mathematical manipulating, and his activities were primarily mathematical discoveries and constructs. These basic mathematical notions were neglected or diluted as childhood development schooling became more formalized. For instance, American psychologist Edward Thorndike defined learning as connections reinforced or diminished by outcomes, including such incentives in the first years of the 20th century. His recommendations for schooling were to go straight to the point. He advised replacing the first Froebelian gift (tiny spheres) with a toothbrush and the first occupation with "nap" to promote wellness. As a result, the gifts' mathematical base was overlooked. People and pets are both learners. Animals attend schools and are students, for instance, at a dog training academy (Mishra, 2007).

Curriculum management and design, as per Hosford in Bamus (2002), includes

- Consensus on the task's essential parts.
- Definition of goals and purpose content Marketing selection
- Organization and sequencing of educational chances
- Material and facility selection
- Assessments

In Akangbou (1984), Lewis and Miel define curriculum as "a collection of ideas regarding possibilities for humans to be trained with other people and objects (all bearers of knowledge, procedures, methods, and values) in specified patterns of time and space."

5.2 Curriculum Design

Curriculum design, as per Taba (1962), is "a declaration that recognizes the components of the syllabus, states their relations to one another, and suggests the rules of organization and the criteria of that organization for the organizational situations under which it is to operate." A curriculum concept that sets the resources to analyze and the principles to implement must be supported and made apparent in a course design.

The essential term in Taba's (1962) remark about curriculum planning is organization. I believe that such a reference to organization reflects the human element of this approach by acknowledging that it was an organizational and judgment procedure that eventually influenced the growth. Design is concerned with deciding what should be studied, how it should be studied, and why it should be learned. The New Zealand Curriculum (2007) demonstrates this: Curriculum development and evaluation was a cyclical procedure. That entails creating choices about how to implement the school curriculum in methods that best fit the wants, passions, and conditions of the students and teachers at the school.

According to Ornstein and Hunkins (2009), the curriculum design should encompass the transmission of basic concepts, behaviors, and abilities to students. As a result, curriculum design considers intellectual growth and characterization, and the connection to curriculum design is tied to specific curricular philosophy often favored by people participating in the design and development phase.

5.2.1 Major Elements of Curriculum Development

Ornstein and Hunkins (2009) distinguish two primary parts of curriculum planning: the resource of the development, such as what resources inform curriculum development, and the actual technique or technicality guiding the planning process, often known as a lateral and vertical organizational influence. The curriculum development can be influenced by several factors, such as social implications, moral ideology, skills, and the student (Ornstein and Hunkins, 2009). When creating a curriculum, the resources could be combined to include modern and classic philosophies. The source of concepts is culture, which relies on economic, cultural, and political settings and incentives. There is a strong emphasis on meeting the demands of today's and tomorrow's society. Information as a resource prioritizes what understanding had the most significant worth and could be conveyed to the entire community. The planning decision is affected by identifying what knowledge is fundamental to society. The learner as a resource of curriculum development complements information as a resource of curriculum planning since it concentrates on the requirements of students and how they study and improve attitudes, ethics, activities, and personality. This notion is intended to notify curriculum design, suggesting that student-centered learning and teaching strategies are required.

The lateral and vertical structure of the curriculum is the second part of curriculum planning. The vertical organization guarantees that the curriculum is organized logically and consistently. It applies to the vertical organization of material in the curriculum, as evidenced by sequencing, consistency, and vertical articulating or integrating (Ornstein and Hunkins, 2009). The series is how material (such as principles and techniques) is presented across time, whether by week or year, to maintain continuity. The horizontal organization is primarily concerned with the scope or the depth and breadth of information given (Tyler, 1949). Subjects, concepts, ideas, philosophies, theories, and pedagogical approaches were all included in the scope. The project also includes learning and teaching objectives and also competencies. Furthermore, the horizontal organization is focused on side-by-side linkages or implementing various curriculum topics. For instance, a social studies topic might be handled in a literature class.

Coherence and consistency are other critical aspects of curriculum planning, in which all portions of the syllabus have to be logically coherent. All elements should fit together, or, as Ornstein and Hunkins (2009) put it, there should be an equilibrium where abilities, behaviors, and skills are linked with ideologies. Every structure element should be given proper weight in the equilibrium (Ornstein and Hunkins, 2009).

5.3 The Processes of Planning and Development of Curriculum

The procedure of curriculum management and designing consists of 5 phases of operations. The phases involved are objective setting, procedures (Technology), the applications phase, execution, and assessment (Longe, 1984).

5.3.1 Goal Determination (Aims and Objectives)

This duty entails determining the overall educational aims and describing the key goals. The overall goals of education were political. They were frequently articulated in general words to get the support of the bulk of the population. They act as a base for determining how school experience could be organized and what should be required to teach, but they do not immediately dictate or define the applied elements of school life (Longe, 1984). Education is a life orientation, seeking to "raise the availability of high workforce," or promote a "more complicated understanding in children," to name a few political goals. Government or legislative measures officially declare the goals of education in general. Three main aspects influence curriculum aim choices. The student, community, and subject matter are all involved (Akangbon, 1984; Bamus, 2002)

5.3.2 The Society

Education organizations are influenced and demanded by demands, beliefs, and other social influences. The world around us is constantly changing. As a consequence, the program planner should choose teaching goals that take into account these changes. When determining a new course of study for the school, the curricular designer must consider the ramifications of societal developments. Employment patterns, as schools have to teach essential abilities, and the needs coming from the emergence of new performance in health, benefit, and political work are among the social changes to address. Incorporating ICT, HIV/AIDS education, and measurable skills in the Basic Education curriculum is a good example (FME, 2007).

5.3.3 The Needs of the Learner

This comprises the learner's traits as well as how he studies. The learner's career and professional demands must be recognized in curriculum preparation, just as they must be throughout society. It is because education determines to create people worthy of living and to live with.

5.3.4 Subject Matter

We will look at particular areas of knowledge, such as nature and philosophy, in this section. The topic material is constantly changing, just as culture is. In this respect, the curriculum that will be developed must take into account the implementation of new conclusions, subjects, and structures that are essential to living in today's world.

The primary educational goals (MEO) in them and achieved via their studies must generally be defined when the education program is organized as per subject material, with specific curricular materials prepared for every subject. The MEO defines broad categories of human behavior and should be turned into more precise rules for creating instructional resources first (Akangbou, 1984).

The redesigned 9 National Mathematics Curriculum for primary education in Nigeria, for example, aims to provide learners with the ability to:

- Develop the mathematical thinking needed to function in the information era;
- Develop the comprehension and application of statistical abilities and ideas required to thrive in an ever-changing technology world;
- Create the critical component of solving problems, interaction, logic, and ability to connect required to thrive in an ever-changing technical era;
- Recognize the main mathematical concepts, considering that the world has transformed and continues to change since the original National Mathematics Curriculum was published in 1977(FME, 2006).

Even though the three components listed above are the most important in determining educational curriculum goals, the curricula designer must also examine the features of the education systems before making decisions. This could be defined as the education system's total academic objectives and organizational characteristics.

5.4 Curriculum Development

The new curriculum could be described as a series of phases that lead to creating a complete curriculum. That entire syllabus is a living document that must be revised and examined regularly. The new curriculum, according to Sahlberg (2006), is defined as:

Furthermore, curriculum design is no longer a project with a beginning and a conclusion. In today's rapidly evolving world, a curriculum developed now and applied in the future may still be relevant and valuable theoretically in 5 years, but concrete details may not be. Pedagogy must be considered as a living, natural tool that aids instructors and institutions in determining the best methods for educating kids.

The curriculum was often in a phase of constant adjustment since it was sensitive to ideas and evolving requirements of society, according to the Education Resources Information Center (ERIC). It is underlined by Braslavsky, who believes that curriculum design should be a continual activity. Usually, the planning process involves numerous stages, each affected by the perspectives of the people concerned. According to Braslavsky, one instance of such a method is a four method to curriculum building that looks like this:

- What culture or families want;
- Educator replies in school;
- The collecting of these answers and the attempt to discover any common elements; and
- The creation and assessment of shared standards.

Marsh and Willis (2007) distinguish national, state, provincial, school, and class stages of curriculum design, each relevant to this study. At the national scale, it is vital to remember that the national curriculum boards at every level are made up of various people, and the instructional design procedure might be assigned to more than one level, resulting in a networked instructional design process. The composition of the curriculum creation team is determined by whether the procedure is general, high-level, or location. "A principal, an assistant principal for curriculum, a group leader, ahead of the department, or by directing class instructors," for instance, at a site level (Wiles 2008). Curriculum design, in essence, is the process's output aspect.

5.4.1 International Approaches in the Curriculum

Most curricular papers describe the school math curricula as a collection of 'process' and a collection of subjects or' content.' The essential notions of mathematics are usually covered in the content, traditionally divided into themes including numbers, arithmetic, measurements, geometrics, probability, and data. Although procedures comprise the behaviors related to using and using math to resolve issues that may be routine or non-routine, they were bundled together and labeled Working Arithmetically in many state and territory mathematics curriculum materials (Clark et al., 2007). The preceding section highlights the problem-solving approach in the math curriculum and the assistance support for educators in Singapore, England, Hong Kong, and the Netherlands. These nations were selected to showcase some of the methods taken and emphasize implementation challenges.

5.4.1.1 Hong Kong

Wardlaw (2008) stated at a National Curriculum Board event that Hong Kong had experienced substantial reform since 2000, emphasizing learning outcomes via curriculum, pedagogy, and evaluation conformity. The following essential concepts are linked to this reform:

- Life-long learning skills will be required for a modern and prospective world;
- Entire user advancement for improving the quality of life in values, heritage, and economic system; adequate knowledge conceptions were also shifting – cross-disciplinary, individual, and co-constructed; and
- Fundamental changes enable possibilities and paths for all youngsters (Wardlaw, 2008).

5.4.1.2 Singapore

The findings of an earlier TIMSS research in Singapore resulted in various adjustments in the curricula, including a 30 percent reduction in material (Kaur, 2001) and solving problems becoming the main objective of studying mathematics. Problem-solving is based on five interrelated parts of the math curriculum: abilities, ideas, procedures, attitude, and metacognition. Although attitudes indicate the numerous factors of education, metacognition emphasizes the significance of self-regulation, and processes encompass gaining and using math skills; the chapter explains in detail tools and abilities.

5.4.1.3 England

In England, the most recent mathematics curriculum documents for Key Phase three and Key Stage 4 (the first four years of secondary education) were even less prescriptive, allowing educators more latitude. They include a structure for individual thinking and learning abilities and an emphasis on learning evaluation. Problem-solving, regarded as "lying at the core of math" (DCSF, 2008), is portrayed as a cycle of activities that includes describing, analyzing, analyzing, and evaluating, as well as sharing and evaluating. "The graphic depicts the dual character of mathematics: that was both a solution for addressing issues in a range of situations and a subject with a unique and logical structure," says the explanation for the connections (DCSF, 2008).

5.4.1.4 The Netherlands

Researchers at the Freudenthal Institute in the Netherlands have worked on a math curriculum, and teaching method called Realistic Math Education for at least 30 years (RME). The concept is built on the idea that math is a human influence and that students should have the opportunity to practice inventing or' mathematizing' during sessions. Problem analysis of possible contexts (those that make sense to students) has been used to educate math skills and behaviors. Instead of adopting a more typical teaching technique of demonstrating basic mathematics, practicing skills, and then putting them into situations, this process begins with real-world challenges to develop learning mathematical concepts. Nevertheless, because the emphasis was on problem contexts that are 'conceivable' or' realizable' for the learner, a more formalized issue may be suitable for such pupils (Van den Heuvel-Panhuizen, 2003). A few other nations, such as the United States and England, have adopted the philosophical perspective established in the Netherlands (Romberg, 2001).

5.4.1.5 The Australian

Mathematics is presently being developed, with consultations beginning in early 2010 and execution beginning in 2011. The Configuration of the Australian Curriculum: Math, which serves as a guidance assertion for writers, divides the configuration into three content strands –arithmetic and number, evaluation and geometry, and probability and statistics – and four proficiency strands – understanding, language skills, solving problems, and reasoning (Kilpatrick et al., 2001). "The capacity to make, understand, conceive, model, and research issue situations, and communicate effective solutions" is how to issue solving is defined. Teachers will require practice models to promote smooth execution; therefore, expectations for problem-solving would be expanded to assist valuation practices.

5.5 Models of Curriculum Development

The following are a few examples of curriculum development designs:

5.5.1 Tyler's Model

Tyler's study in Bamus (2002) was centered on four key questions:

- What academic objectives should the school strive for?
- What educational opportunities exist which are likely to achieve these goals?
- What are the best ways to organize these education opportunities?
- How can we tell if these objectives have been met?

Tyler makes no effort to address questions; instead, he contends that the answers will differ depending on the level of knowledge, and he recommends techniques for analyzing these techniques, which he diagrams below:

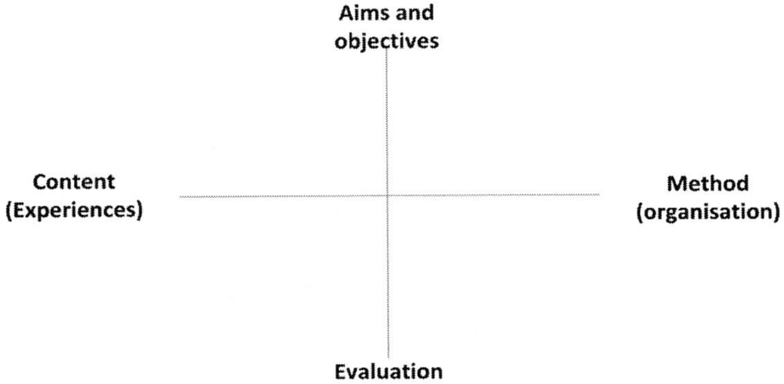

Figure.5.1 Tyler's Model

5.5.2 Wheeler's Model

Few contend that the Tyler model is overly simplistic and that review should occur at every stage rather than only at the end. Per Wheeler in Badmus, Tyler's original concepts were converted into a cyclic structure with five steps (2002).

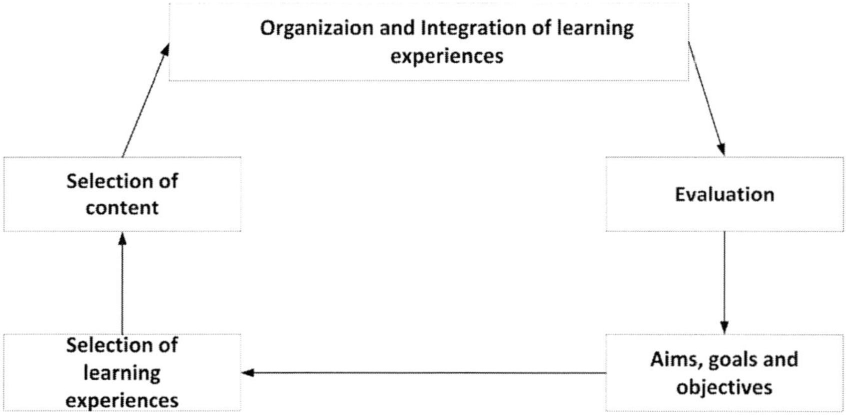

Figure.5.2 Wheeler's Model

5.6 Evaluation of Curriculum

The 4th step in curriculum development is evaluation (Ornstein and Hunkins, 2009). Curriculum aims, curriculum goals, curriculum structure and execution, and curriculum assessment are the four parts of Oliva's (1997) curriculum design. The main objective of program evaluation, as per Oliva (1997), is to evaluate whether or not the curriculum aims and goals are now being met. School performance should not be utilized to gauge a curriculum's efficacy or effectiveness if the project's aims or goals do not include it (Oliva, 1997).

Hamilton (1977) pioneered curriculum assessment, focusing on academic performance and whether or not pupils are achieving them. This technique has persisted until today, with empirical methodologies used in curriculum design. The international use of international testing valuations, such as the Program for International Student Assessment-PISA and The Trends in International Mathematics and Science Study, according to Malik (2010), is among the causes behind this TIMMS. Mathematics curricula in some nations like Finland and New Zealand are PISA-compliant. According to Tienken (2013), the findings of the PISA and TIMMS exams do not deliver an evaluation of the worth of public education in the United States because of the small population of youth tested in the US compared to other nations.

According to Scriven (1967), there are two kinds of curriculum design. The first was summative, starting at the curriculum phase of development so that when the curriculum is created, it is also reviewed and focused on the questions raised in the preceding two sections. Formative assessment is inferred by its name to be a continuing process that connects it to the continuous aspect of the curriculum creation procedure. The second form of program evaluation outlined by Scriven (1967) happens after the established curriculum is used to evaluate or assess it against the curriculum goals.

5.7 Pedagogy of Mathematics

According to teachers, some techniques will be more suitable to their local conditions than others. The fundamentals of math education as a whole must:

- Admit that all students, regardless of age, could promote better mathematical personalities and become efficient numerical learners;
- Be based on interpersonal care and respect, and be attentive to the multitude of cultural heritages, thought patterns, and facts typically fouled by mathematics;
- Be dedicated to improving a range of social results in the classroom context that will focus on students' overall development as productive members of society, such as fundamental understanding, procedural knowledge, thinking skills, and regulation of cognition; and
- Be able to focus on optimizing a variety of attractive learning achievements, such as fundamental understanding, procedural knowledge, thinking skills, and regulation of cognition.

Teachers' knowledge of and capacity to use and assess educational resources and materials, such as new tech; ways to reflect mathematics concepts and processes; teaching approaches and classroom organizational structures; help promote debate and build a sense of mathematical society; and methods for evaluating student knowledge of math should be developed during preservice and ongoing training for teacher educators.

5.8 Pedagogy of Mathematics Teaching

Maths pedagogical is concerned with how teachers assist their pupils in taking the time to understand, perform, and apply mathematics. This guideline specifies various pedagogical elements that are critical to good teaching. These elements serve as filters through which instructors can filter their concepts of math and pupils to benefit and improve their math instruction.

Teachers are all in charge of assigning worthwhile mathematics tasks to their students. They can use pre-made projects or create their own to keep pupils' learning abilities on the track. They frequently use a range of study materials and tools, such as issue booklets, concrete elements, books, software applications, calculators, and so on. Teachers require a very well structure for finding and evaluating learning material and technology devices and knowing how to employ these sources in their classrooms successfully.

Teachers who sincerely care about their students work hard to ensure a secure educational environment. They also make some that their classes have a strong math emphasis and an excellent but practical assessment of their student's capabilities. Learners determine they can think, explain, express, replicate, and critique the math they experience in this environment, and their learning environments become a source for upward their math talents and identity.

Students require work opportunities individually and collectively when trying to make sense of ideas. They may need to reflect and work in peace at times, away from the demands of the entire class. They may need to work in groups or pairs to exchange knowledge and thoughts. At some other times, they must be active participants in a meaningful, whole-class conversation in which they can confirm their knowledge and be introduced to broader perspectives of the mathematical topics under conversation.

Teacher educators put individuals' current knowledge and passions at the forefront of their educational decision-making when preparing for learning. Rather than attempting to correct flaws and fill gaps, they build on the existing skills and adapt their training to match the requirements of their students. They can use their learners' thoughts as a source of more knowledge because they see thought as "knowledge in development." These teachers know of their pupils' needs and the subject of mathematics.

In work to explain a new notion or capability, students should be able to connect it to their previous math conceptions in a variety of ways. Learners could grasp the connectivity of diverse mathematics concepts and the linkages between mathematics and the natural world by completing tasks that require them to establish links within and throughout topics. If students receive chances to use mathematics in real-life situations, they understand its benefit to society and contribute to other areas of expertise, and they begin to see math as an aspect of their individual experiences and lives (Ainley et al., 2006).

Excellent educators support pupils in using and comprehending language widely accepted in the mathematical community. They accomplish this by establishing linkage among mathematical language, instinctive knowledge and understanding, and the student's native tongue. Ideas and specialized phrases should be taught and modeled in terms that students may know while remaining loyal to the deeper meanings. Teachers can make pupils conscious of the variances and the intricacies in mathematical terminology by correctly discriminating terms.

5.9 Summary

This chapter explains what curriculum design is and what the primary parts of curriculum planning are. Any nation's academic program aims to produce idealistic men and women in civilization. Any flaw in curriculum management and design has ramifications across the educational system. As a result, curriculum designers must guarantee that the curricular phases were followed. Among them are goal setting, objective selection, organization, tryouts, implementation, and evaluation. Curriculum design is fluid and adapts to the changing needs of society.

CHAPTER 6

TEACHING AND LEARNING MATHS IN SCHOOL

6.1 Introduction

Mathematics is, without a doubt, the most necessary course of study. On the other hand, mathematics rarely seems to have the same daily potential to affect a pupil and can usually only be learned through the education system. Teenage kids learn mathematics in school to prepare themselves for a career beyond school. Such knowledge can be used in a variety of ways. It might be about gaining knowledge to be a more educated person. It might be about understanding better how mathematics performs an essential yet underappreciated part of 21st-century living. It might be about preparing for the workplace. It can be regarding post-secondary schooling. It could also be for knowing how to live in the real world. Various mathematical activities, multiple methods of learning mathematics, and unique approaches to becoming motivated through mathematics are all achievable goals. Whichever the case may be, each young person in the country must study maths in schools. For many, though, it is a challenging and ultimately detaching situation. Student learning in mathematics is a vast and active discipline focusing on psychology, sociology, cognitive science, and philosophy. (Sutherland et al. 2004)

6.1.1 Teaching and Learning

A primary goal in education tends to overlook the role of a teacher, whereas a sole concentration on educating overlooks the variability of pupils' knowledge. Consider an individual learner as a means of learning why they learn math. Particular student provides an example for educators to realize how these students learn, in a comparable pattern that learners are provided mathematical examples to help learners comprehend mathematics. Education can be described as the transition from not knowing to knowing something. Students study most mathematics in class, and in most classes, professors lead students through a portion of the syllabus. The mathematics syllabus is constructed so that each session builds on previous math that the pupils have learned so that a good session will progress to more mathematics.

The educator can incorporate multiple instructional methods to provide different instructional approaches. His or her choices of learning method in each session are based on previous practice with learners and might be impacted by the school's prior undergraduate years (Dean 2019). When educating pupils, it has been observed that every issue seems to be more certain to be appreciated if the learners receive a holistic understanding of the situation. Teachers reviewing the learning program at their school are more likely to learn the intricacies if they are viewed from the perspective of the education process and its historical evolution.

6.1.2 Mathematics as a Human Activity

School mathematics must be understood as a human activity representing mathematicians' efforts to understand how some procedures work, explore different approaches, explain statements, etc. It should also show ways mathematicians analyze a complex problem, choose factors, determine how to define and integrate the factors, perform operations, predict outcomes, and check the predictive performance. The group of mathematicians, mathematics teachers, and math students defines the field of mathematics as a vast aggregate of concepts in various linked subject areas. Mathematics is indeed a subject of comprehension and a method of learning. Structures and languages were critical to comprehending the physical, social, and mathematical worlds people encounter. Mathematics should be firmly entrenched in and related to the experiences to support students' requirements to develop a sense of experience beyond the student learning and mathematics itself, especially thinking clearly in several sciences (Romberg 1999). The chapter is intended to provide school students and teachers with a broad understanding of mathematics teaching and learning.

6.2 Students Engagement in Learning Maths

The concept inspires some educators that education may give kids chances that seem free of their home circumstances or, at the very least, aim to mitigate the detrimental effects of specific cultural traits. However, it appears that the pupils who stand to benefit far more from class participation are also the ones who are the most challenging to motivate, with such an issue peaking in higher middle and secondary school. Also, in schools that are attempting to take substantial steps to address the importance of the content, substance, and type of work established, the significance of analysis and reporting, and the inclusion of students through the adaptation of instructional practices and techniques of assembling, discussions with junior academic staff, particularly those representing lower socioeconomic populations, prove the above lack of engagement.

Various circumstances can cause a lack of school engagement and positive educational achievements. It is likely that students in the intermediate grades lose confidence, or that they need abilities, that they stop trying, that students do not even see the value of mathematics, or because they are unaware of actual challenges, and that they believe they can succeed in school despite putting in any work. To go deeper into these issues, researchers looked into individual students' opinions of how much their attempts influence their academic achievement and enthusiasm in general and mathematics in particular.

6.2.1 The Pupils' sense of Self-Regulation

The study looked into one side of the pupils' ambition, specifically their ability to control their actions. Initially, it is viewed as something that every student must decide as well as responding of their own. The study uses a unique approach to research focusing on educators' ways to tackle student motivation, and it suggests a unique viewpoint to research that examines or explores feelings, beliefs, or values. Perhaps one hypothesis is that the student's motivation is a factor in their failure to engage. The actions that students make are influenced by their motivation. Of all, results can be achieved to act or not act, based on need or fear, and can be aimed at specific requirements. Needs could be classified as focusing on self, cognitive functioning, or social interactions and can direct specific behaviors.

Identity, autonomy, and consistency are all aspects of personality. Socioeconomic referring necessitates good reactions from others and a desire to help them. Cognitive performance necessitates sharing information, which includes understanding oneself, everyone else, and the world. While learning can trigger all of these requirements, it is not always practical, and children may meet the demand in ways that are contrary to the school's and teacher's intentions. Objectives were later considered a particular subject of need by the experts. A pupil with a yearning for freedom and an underlying purpose of undermining the teacher's power was given as an example. Researchers were concerned about how the pupils achieved their particular goals and their ability to think about their aims. Researchers wanted to understand more about the factors that may affect the character of pupils' needs, goals, and, ultimately, the choices they make regarding their involvement in mathematics learning.

6.2.2 Possible sources of the lack of Pupil Engagement

Researchers looked at the pupils' self-regulation attitude and ability to better comprehend the reasons behind their choices. The basic methodology was built based on the findings of a specialist who differentiated between two approaches to cognition. One is entity theory, a static view of intellect in which individuals assume their ability is preset at development and remains constant throughout their lives. According to the researcher, pupils who believe in the entity perspective need easy wins to stay motivated and regard problems as risks. Students who believe aptitude is changeable or progressives believe they can modify their ability and achievement by adjusting variables over which they have limited responsibility.

The approaches in which students identify their needs and goals are inextricably linked to such ideas of knowledge. When students have difficulty, the analysis states that they lose faith in themself, have fewer expectations, develop negative mindsets, have less tenacity, and have higher levels of depression. Such pupils seek out positive feedback from others and reject negative feedback. Students with such ambitions are more likely to be resilient in the face of loss and to keep concerned with obtaining skills and information when faced with difficulties. They do not accuse someone else of threats and do not perceive defeat as a personal failure; instead, they set learning objectives to improve their skills when faced with adversity. Students who view intellect as progressive and do not require accomplishment to achieve learning tasks have no confidence and self-esteem (Sullivan, Tobias, and McDonough 2006).

6.3 Teachers' Beliefs about Teaching and Learning Mathematics

The non-alignment of language employed by mathematics education academics has been a roadblock to advancement within that field. The expressive domain is typically recognized as relating to structures that go beyond the cognitive domain. Beliefs, attitudes, and emotions may be classified as subtypes of impact, regardless of the lack of consistency between mathematics education experts on vocabulary. The attitudes of mathematics teachers influence their classroom practice, how they view teaching, learning, and evaluation, and how they view pupils' perspectives, skills, attitudes, and talents. Researchers examined the attitudes of primary mathematics teachers who participated in a mentoring program toward the beliefs of their peers, which was before teachers who did not participate in a mentorship program (Tasos) (Barkatsas and Malone 2005).

Educators in the intervention group engaged in small-group instructional activities of the mentoring intervention based on a social-economic strategy for mathematics instruction. It was discovered that the study participants behaved as though the mentoring program's socio-constructivist method of mathematics instruction had first affected their attitudes. However, their activities later in the program revealed that they supported a more conventional set of attitudes regarding mathematics education. These researchers appreciated past academics' efforts in the development of the concept of reflecting and provided the applied framework equally for describing instructors' knowledge belief structures:

- *Isolationist:* Teachers who come into this group have cognitive frameworks that keep their beliefs segregated or grouped aside from others. Isolationists are not known for their willingness to accommodate others.
- *Naïve idealist:* Teachers within the group are accepted knowers in that, opposite to isolationists, they accept what others perceive to be true while rarely questioning what they think.
- *Naïve connectionist:* In contrast to one's views, the position encourages introspection and attentiveness to others' opinions. On the other hand, naive communication and social skills cannot resolve conflict or belief disagreements.
- *Reflective connectionist:* It emphasizes introspection and interest in other people's views rather than oneself. Reflective connectionists, on the other hand, use introspective thought to settle disagreements.

The researchers suggested that school and classroom procedures, rather than teachers' beliefs, were the primary sources of the contradictions revealed in the analysis. The main reason for contradictions among beliefs and practice was the combined result of various external variables (i.e., previous school events, teaching educational intervention, character features of the instructor, social, educational standards, etc.). Teachers' beliefs in mathematics classrooms are prone to be affected by: (a) the teacher's previous observations with mathematics, teaching skills, interaction with peers, the consciousness of mathematics education research findings, using mathematics in other personal experiences, and teacher education programs; and (b) wide variety of individuals, economic, cultural, and interpersonal factors.

6.4 Using Technologies in Teaching and Learning Mathematics

Students can visualize mathematical ideas which are hard to picture through standard instructional techniques because of multimedia features. The primary forms of digital learning approaches adopted in teaching and learning mathematics are the Computer Algebra System (CAS), dynamic geometry software, and spreadsheets. Differing programs, on the other hand, allow instruction at various curricular levels, with various amounts of time required for learners to become competent with the program. The use of computers in teaching and learning improves not just student achievement but also enthusiasm. Information and communication technology (ICT) are the main determinants in the information age. According to the report, 'An examination of the meta-analyses of computers in education reveals that computers are used efficiently,

(a) when various instructional methods are used.
(b) when pre-training in using technology as teaching and learning aids is provided.
(c) when there are numerous studying chances.
(d) when the learner rather than the educator control knowledge.
(e) when peer tutoring is at its best.
(f) when the response is perfected.

6.4.1 Teaching and Learning with Mobile Devices

While it is true that using technology in education can promote modernist techniques, it is also true that using technology does not necessitate a dramatic shift in education. Studying through technologies entails more than simply converting actions to digital formats; it also entails the construction of "authentic learning environments" that create important use of new ways to improve knowledge acquisition, interaction, and presentation. The primary duty for integrating technology into the school rests with the educator, who should encourage such an educational revolution. Three critical aspects of instructional tools have been identified:

1. the potential for adapting content
2. the potential for innovative instructional methodologies
3. the potential for a shift in attitudes

Many academics emphasize the importance of investigating perspectives, emphasizing that teachers' views about creativity are essential to the achievement of implementing new technology into teaching practices (Polydoros 2021).

Aside from the necessity of assessing teachers' perspectives on integrating smartphones into the education process, there is a lack of research on students' perspectives on smartphone usage. According to evidence on the effects of portable devices on learning, they can benefit students by establishing an environment in which they may develop and share information and have superior learning results. Furthermore, the findings demonstrate that using smartphones improves students' interest and enhances compelling learning experiences.

The leading smartphone practices in the school setting were discovered through research. They were searching online (51%), emailing (46%), reading novels (42%), making notes (40%), completing assignments (30%), reading newspapers (26%), watching youtube (26%), online communication (25%), texting (23%), buying stuff (15%), and generating lectures or files (15%) and the most school tasks (12 percent). As a result, all students and teachers identified two kinds of instructors due to the mobile use method. Furthermore, the findings revealed that teachers' attitudes toward smartphones affect their teaching techniques.

6.4.2 Teaching and Learning Maths through Games

Students establish a feeling of mathematical concepts while playing games and participating in arithmetic tasks with their peers and then operate on it by developing and revising their ideas. Creating a consistent and organizationally maintained instructional framework could be a cornerstone to improving teachers' competence. The education system relies heavily on instructional methods and the creation of educational facilities like "teaching and learning using games and activities." The percentage of students submitted a list of games and exercises with responses (minimum of ten instances). Students spent hours online and in textbooks looking for these instances. Every field of electrical and computing student is required to study mathematics. As a result, they will be more productive in their profession if they are educated using approaches that draw theoretical mathematics concepts closer to their understanding.

One technique is to encourage students to code games, while the other is to seek up other games. Their inventiveness, rational reasoning, and problem-solving skills could all benefit from it. Every learner got his or her computer, and the instructor had a series of complaints written on the chalkboard. The pupils subsequently worked in groups to solve the difficulties. Students collaborated in this way, and the role of the teacher is limited to assisting them with the task and then discussing their respective conclusions. Students were also given assignments to finish tasks comprising multiple mathematical problems and were permitted to play with mathematics software systems (Gyöngyösi Wiersum 2012).

6.4.2.1 Hints for Successful Classroom Games

- Double-check that the game corresponds to the mathematical goal;
- Don't merely play games to pass the time; use them for a purposeful goal.
- Reduce the number of gamers to two to four to ensure that rounds are completed swiftly;
- There should be enough element of luck in the game for slow learners to believe they have the potential to win.
- Keep the game's finishing time to a minimum;
- Choose five or six 'basic' activity formats to get the kids used to the principles-change the math instead of the restrictions;
- For schoolwork, send a child back with a well-known game;
- Encourage kids to design their board games or variants on well-known games.

6.5 Elementary Students Learning Maths

In elementary school, kids initially encounter mathematics successes and failures. As a result, elementary education sets the foundations for pupils' mathematical abilities. Teachers have a more significant impact on pupils' mathematical achievement in elementary school than in subsequent years. Elementary school teachers expose learners to mathematical terms, symbols, and mindsets. The core abilities pupils need in the subsequent educational career are taught in this beginning. For the scenario mentioned above to occur successfully, elementary school instructors must use engaging, inspiring, and inspiring mathematical projects and instructional activities.

Thus, through mathematical experiences in elementary school, students might consider mathematics entertaining and beneficial, and they can acquire good perception by appreciating mathematics. Furthermore, elementary school teachers must learn basic settings that encourage students' engagement in mathematics. Learning systems in which students can articulate their ideas and communicate them with their peers and professors give students the possibility to create solid mathematical concepts, significant ideas, findings, and new information can be presented. Students' numerical enthusiasm and personality are enhanced in such educational contexts.

6.5.1 Students Learning in Number form

According to the findings, numeral-number-four operational indicators were used the most in illustrations at all grades regarding numbers and symbols in the educational setting. The foundation of primary school maths is the instruction of numbers and operations. The procedure of arithmetic, deduction, multiplying, and dividing will make up the majority of elementary mathematic instructions. Numbers are at the core of every mathematics teaching around the world. Young children's first mathematical thinking is based on numbers, probably their earliest mathematical expression. In the elementary school mathematics program, geometric, measuring, and data analysis learning areas must not be overlooked (Turgut and Turgut 2020).

6.6 Mathematical Problem-Solving of Middle School Students with Learning Disabilities

Students comprising three stages of problem-solving capacity (i.e., learning disability, average-achieving, and blessed) solve mathematical problems in grade three stages (i.e., six, seven, and eight) to generate insights into data structures, ideas that help, and procedural knowledge of students, especially those with learning difficulties. In academic disciplines such as reading, writing, mathematical calculation, and mathematics problem solving, children without learning difficulties use cognitive and metacognitive methods in a limited manner. While some children may lack essential and core critical, others may have this knowledge but do not apply it in the same context or reaction to the same indications as typically growing thinking skills.

Analyses that relate communities of varying ability and proficiency levels are essential to understand the cognitive, metacognitive, and affective variables that communicate and impact numerical problem solving and separate particular inadequacies that interact with proper and effectively solving problems. The study mentioned the value of investigating extremes on an assumed range of problem-solving skills in a debate on gender variation in arithmetic problem solvers. The dearth of knowledge about variations in problem-solving abilities between relatively competent or ordinary pupils and other ability groups has been noticed. It suggested researching children with different levels of mathematical problem-solving competency better to understand the performance levels within this functional area and to facilitate education that promotes or improves transition among levels. (Montague and Applegate 1993).

Affective characteristics such as low motivation and poor consciousness affect solving problems in addition to cognitive and metacognitive factors. Students with learning disabilities exhibited significantly worse sense of self than nondisabled comparison group or comprehensive norms, according to a review of literature on consciousness and intellectual difficulties. The component is essential for examining problem-solving activities because persons with a favorable identity of performance attempt more fantastically and endure a lengthier, more challenging, or more demanding activity. The current research looks at various cognitive, metacognitive, and affective characteristics that are thought to be required for proper and effective mathematical problem-solving.

6.6.1 Cognitive Attributes

It entails procedural memory, mathematical understanding, the opportunity to utilize that information for verbal problems, and the understanding and application of two critical problem-solving methods: problem presentation and issue solution. Problem representations entail converting language and numerics into suitable mathematical formulas and processes and using logical, integrative problem frameworks or representations that are verbal, graphical, conceptual, and numerical. Problem-solving entails planning and scheduling techniques (i.e., identifying alternate and uncommon answers and approaches) (i.e., implementing a working-forward strategy instead of a trial-and-error or means-end approach). Though several kids with learning difficulties have general problem-solving skills, the most significant shortfall appears in tools and procedures primarily related to the problem description, according to the mathematical problem solving with middle school pupils. Rephrasing or reiterating problems in one's terms, visualizing problems using pictures on paper or mentally picturing, and hypothesizing, which entails formulating objectives and devising methods to address problems, are some of these methods. Problem representational strategies are necessary to analyze verbal and mathematical information, build internal representations in mind, interpret and incorporate problem knowledge, and generate answer ideas.

6.6.2 Metacognitive Attributes

It differs from cognitive traits in that it emphasizes cognitive understanding of personality, the deployments or application of mental abilities or approaches in solving problems, and management across techniques for the goals of organizing and coordinating output. The capacities to construct assumptions about problem solutions, continuously analyze the solution path, and analyze outcomes are all examples of metacognitive traits.

6.6.3 Affective Attributes

A favorable mindset regarding mathematics and problem solving, curiosity in solving problems, educational freedom, and faith in one's capacity to solve problems are all linked to improved mathematical problem-solving. Behavior regarding maths and problem solving, self-assessment of competence, and assessment of the relevance of mathematical problem solving where all the study's emotional variables. A systematic approach, including simultaneous and prospective verbal reports of problem-solving activity, was employed to investigate these cognitive and non-cognitive qualities. Informal and unstructured conversations, able-to-think tasks, and focus group tasks are used to acquire descriptive information that can be examined subjectively or statistically using the technique suggested for cognitive process examinations of solving problems. Despite the current trend of employing this analytical technique to research cognitive and non-cognitive aspects of learning, outcomes from studies based on speech performance reports should be evaluated with caution concerning the measurements' validity.

6.7 Gender Differences in Perception of Teaching and Learning Mathematics

Mathematical beliefs are influenced by students' views of the school setting; academics from around the nation have identified mathematics as a breakthrough subject that divides pupils regardless of their social status or gender. Some study on arithmetic ability has revealed a long-standing gender divide in the advantage of boys. Independent factors have been proposed as causes for the gender imbalance among boys and girls. Girls' inferior arithmetic ability has typically been attributed to both internally and externally situational variables, such as a lack of recognized access to learning maths. Other research linked the females' poor arithmetic ability to student perceptions of their courses as unappealing, unwelcoming, and unfriendly. Teacher and peer support are significant considerations in females' maths success. Overall findings indicate that teachers and peers are favorably related to academic perspectives, performance, temperament, knowledge, ambition, and consciousness.

6.7.1 Gender Differences

It is discovered some discrepancies in how boys and girls interpret their educational environment, as well as variances in how boys and girls view mathematics. Studies discovered that boys believe they use group activities more than girls in different educational settings. Boys generally believe they have more control over the topic and are more engaged in class than females. Teachers begin to socialize more often with boys than with girls, which might support these beliefs. It is performed for various purposes, including lesson planning, where lectures geared at males, typically thought to be more disruptive, serve as a way to keep students on task (Samuelsson and Samuelsson 2016). Girls, on the other side, are more typically thought of as subconscious and on-task, according to teachers' perceptions. Girls who are less engaged in the classrooms or who are never participating usually focus more on their tasks, neither giving nor requesting aid from everyone else. Girls are also thought to have higher arithmetic skills than boys. Girls receive less focus and are less active in-class interactions than boys in a classroom setting based upon the assumption that could be why girls believe they have less control over what is accomplished and which manner is employed. These gender discrepancies may affect how encouraging teachers are seen by boys and girls.

6.7.2 The Classroom Setting and Achievement in Mathematics

Grades are predicted by boys' and girls' opinions of the classroom environment. Student opinions account for 16% of the variance in marks. A supportive organizational atmosphere was the most significant predictor of how generally demonstrates the teaching environment, with a modest but significant meaningful contribution to students' (.25), boys' (.22), and girls' (.25) mathematical performance. These findings are in line with past studies that emphasize the value of a positive social environment. For girls, a supportive setting indicates mathematics performance marginally better than for boys. Students who believe they are involved in decisions about classroom work procedures and what subject must be presented do better than those who do not believe they are involved in such changes.

It was discovered that there is no link between the accessibility of teacher help and higher marks. As an outcome of being allowed increased interaction with teachers, boys' views of someone seeing and acknowledged in the classroom alter their sense of involvement, which has a particular impact or minimum involvement in decision-making. Students' math performance – boys' and girls' – are negatively impacted by an educational environment marked by work overload. The significance of selecting the ideal examples was mentioned in the Swedish National Curriculum in the early 1960s. Before deciding what topic and what should be taught, educators had to incorporate the pupils' prior understanding of the topic. The same ideas resurfaced in the 1980 National Curriculum. Students demonstrated sufficient understanding to begin working in a new field.

6.7.3 Implication for Teacher

Most boys and girls appear to benefit from essential teaching training. Positive elements in mathematics accomplishment include a school environment where students feel a supportive organizational atmosphere, clearly expressed goals, engagement, and fair expectations. The research also reveals how to assist boys and girls in achieving higher results. Because the school environment is a favorable determinant of mathematics performance, teachers should let girls engage more in classroom choices and assist boys in working individually.

6.8 Summary

Mathematics does not have the same daily opportunity to affect a student as other subjects, and it is generally exclusively learned through the education curriculum. As a result, the effectiveness of mathematics at a school directly impacts the students' achievements, as successful learning is dependent on good teaching. It will be demonstrated in this chapter that mathematics teaching and learning can be seen as a gradual system. It also discussed the student's engagement in learning and teacher beliefs in teaching mathematics and using technologies in maths through smartphones and gaming. It also elaborated on how students with learning disabilities can solve mathematics problems and gender differences in their perception of teaching and learning mathematics.

CHAPTER 7

INNOVATION IN MATHEMATICS EDUCATION

7.1 Introduction

Extensive modifications in the area of education have occurred over the last four decades (beginning in the 1970s), and different associations (both government and public and non-governmental) have chosen to take action as a result of or in opposition to the government's policies on schooling and syllabus data provided from time – to – time. The state's current priorities include universal applicability of education and education for democracy. Mathematics has historically been a topic that has resulted in a high percentage of student fails, as well as a cause for students to drop out of school and a source of fear and worry among learners. For many years, various academics have been interested in putting effort into the field to change people's behavior toward and comprehension of mathematics and science that have served as gatekeepers for entrance to higher education. Due to its relevance in the growth of physical and biological sciences and technology, numerous government committees have suggested that mathematics be obligatory for all pupils. As a result, the country has undertaken several attempts in the form of interventions at all levels.

Furthermore, no detailed documentation or research design research has been conducted to acquire evidence of their direct influence on pupils' thinking and learning, training of teachers, or systemic improvements. The current shift in public discourse on general education to a more modern one, incorporating constructivist philosophy and focusing on the child's welfare, has prompted additional efforts in the field. This essay will first explain some of these efforts in the area of mathematics teaching before addressing some of the concerns and obstacles that mathematics research and education in this country confront (Krainer and Konrad, 2008).

7.2 General Objectives of Teaching Mathematics

The main goal of teaching mathematics is to educate students for public life, independent of their job or desired outcomes, on the one hand, and to provide pupils the skills to grasp math themselves, whether in school/after graduation, on the other side. Furthermore, there are other goals to keep in mind, such as teaching kids how to employ good thinking techniques, use inductive and logical thinking, be contemplative and analytical, and learn problem-solving abilities.

Effective teaching is a rewarding job, but it is also a challenging one, and both the enjoyment and the hardship stem from the importance of mathematics and the learner's perspective of it. Teaching, like any other career, involves both knowledge and skill. Teachers must strive to evolve professionally by researching recent innovations in mathematics and participating in events, including conferences, workshops, meetings, and professional training in mathematics, to become aware of the necessary understanding and efficient learning methods and techniques. Teachers can link mathematics to everyday life and present learners with real examples and uses, allowing the user to interact with the educator, the subject, and the educational experience and realize the value of mathematics (Algani, 2018).

Teachers should use imagination when selecting relevant instances for students' daily lives. They should also make connections between mathematics and abstract understood and real-life situations for pupils to recognize and respect mathematics. This method of teaching mathematics aids students' integration into culture and aids their learning of the art of thinking. If mathematics has no relevance to individuals, studying it would be pointless and consist solely of memorization for tests. Teachers should assist students in developing their talents and supply them with all possible means of representation, particularly current ones (Civil and Marta 2007).

Einat Heyd-Metzuyanim (2015; 2016) deduced that the theoretical technique contributes to the growth of mathematics learning among many students and enhances their skills to study maths in her articles about the effect of conceptual and procedural methods on studying mathematics and their correlation with studying designs and the anxiety of mathematics. She also emphasized the significant link between arithmetic challenges and traditional learning methods that tend to dread math and math assessments that she regards as a vicious cycle: Ritual Learning Difficulties in Mathematics Math Anxiety Ritual Learning.

7.3 The Innovations in Mathematics Education Project

InnoMathEd (Innovations in Mathematics Education) is a career growth project including organizations from around Europe. School experiences related to the usage of dynamic mathematics, according to the InnoMathEd proposal, could act as catalysts for learners' active, autonomous, and creative learning in school math. The initiative aimed to use a specific learning environment to promote such learning: the "dynamic worksheet," which uses manipulable constructs and associated text to pose challenges and give solutions.

The InnoMathEd proposal highlighted several critical characteristics of the mathematical creations made feasible by dynamic resources.

They enable visualizations that are not achievable with standard educational material. Unlike those on paper or the blackboard, Structures with dynamic mathematics could be altered and changed on display in real-time. Moreover, incorporated computer algebra systems bridge the distance between geometry, algebra, and calculus. They enable the calculation of lengths, angles, and point coordinates, as well as the application of these data in subsequent calculations. They allow you to interact with functions and integrate function graphs in dynamic structures.

Every project's key partners were connected in some way in delivering initial and in-service teacher education on creating and using fantastic resources in classroom mathematics. Our additional moderate role is to recognize suggestions from the scientific literature on dynamic math that noted significant problems arising at an early stage and then to investigate what forms of use of dynamic mathematics the significant partners had proved to be especially fruitful at a later stage (to be presented and analyzed in later sections).

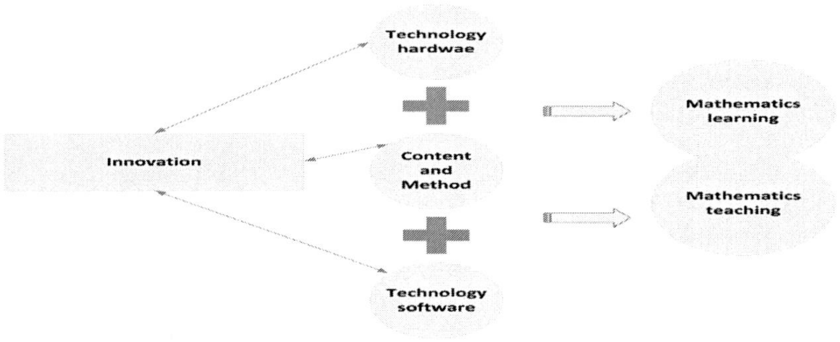

Figure.7.1 Innovation in teaching and learning mathematics

If science and creativity can bring learning and teaching roles closer together with more overlap, it is the area where the greatest profound knowledge of mathematics may emerge. As a result, science and creativity should combine curriculum and instruction to connect learning and teaching for a better procedural and conceptual grasp of mathematics among students.

7.4 Reformations in Mathematics Education over the years

The education system includes mathematics as a required subject. Mathematical abilities are required for various analytic, technical, scientific, and economic applications. In the future, it is critical to teach kids how to utilize mathematics effectively and to recognize its value. Mathematics is required for all to function in an increasingly complicated world. It is not just required for comprehending the other disciplines but also for all to grasp a minimal level of arithmetic (Pech, 2010).

Numerous commissions and committees have worked to improve the quality of teaching in general and math instruction in particular. Mathematical education in schools, both in the elementary and secondary phases, has seen significant reform in recent decades, with new curricula, updated books, and changes to the teaching and learning process. After the National Policy on Education (NPE) of 1968, extensive curricular revisions for all levels of schooling occurred. The New Education Policy of 1986, later updated in 1992, was the next major reform in curriculum and instructional methodologies.

A consistent core element in the school curriculum across the country was advocated in the National Policy on Education (1986). The National Council of Educational Research and Training (NCERT) was also tasked with designing the National Curriculum Framework and revising it regularly, according to the policy. The National Curriculum Framework (NCF) – 2005 is one of four National Curriculum Structures produced by the National Council of Educational Research and Training (NCERT) in 1975, 1988, 2000, and 2005. The suggestions in NCF-2005 are designed to make education a pleasant time for youngsters. It primarily aimed to create a curriculum beyond textbooks and classrooms rather than being bound to them. More focus was placed on activity-based learning and children's general improvement. NCF-2005 also promoted ICT-based learning. In the NCF-2005 guidelines, mathematics lab practices were made mandatory.

Despite widespread approval of the NCF-2005, most mathematics classrooms have remained unchanged because many instructors have failed to transfer the NCF's concepts into classroom practices (National Council for Educational Research and Training, 2005). The National Focus Group on Teaching Mathematics produced a Position Paper in 2006 that included some excellent proposals for improving mathematics education quality (Emron and Dhindsa, 2010).

Though past programs were well-intentioned and well-conceived, they did not produce the anticipated performance in terms of satisfactory educational performance. As a result, the Ministry of Human Resources and Development (MHRD) saw the necessity for a National Education Policy (NEP)-2020. NEP-2020 is a much-needed change, particularly in this digital era, when technology is used in every industry. The focus placed by NEP-2020 on learning systems such as online learning, coding, and digital courses such as Artificial Intelligence, etc., shows that this new policy has considered every step necessary to fulfill the aim of holistic education in order to develop skilled youth for the country (Rajkumar and Hema, 2018). Here are a few examples of student-initiated innovations.

7.4.1 Revival of Mathematical Board Games

It began as a semester-long initiative by a bunch of students to incorporate mathematics into learners' daily tasks through board games and use board games as a teaching method. The inventive and theoretically sound game designs are unique. Each game in the series focuses on one or even more mathematical principles. The complexity and rigor of the ideas raise the game's degree of challenge. The games have a pleasing appearance, simple instructions, and a manageable size. The games offer enough variety to entice players to engage in mathematical research.

7.4.2 Mathematics for Visually Impaired

In most cases, mathematics instruction is geared toward typical students. Students with special needs were usually ignored while creating educational materials. Mathematical concepts relating to space, forms, and measuring are complex for visually impaired youngsters to grasp. Students in the MME program created a Mathematical Kit for Partially Impaired Learners with the help of CIC mentors. Essential geometrical tools like a scale, protector, and compass are included in the kids' one-of-a-kind kit. It also has a specifically designed drawing board for visually impaired children to use to create geometrical shapes. Abacus with shapes to educate number and place value system, tactile board to grasp coordinate system, and block layout to educate numbers with distinct qualities are all included in the kit.

7.4.3 Radio as a means of communication to teach Mathematics

Radio is the most basic and straightforward mode of communication that individuals use in various scenarios. MME students used their knowledge of mass media and communication to launch an on-air radio show to promote mathematics among some of the general public. Students described mathematics themes that would be appealing to the general public. They conducted an extensive study to prepare the material for each episode, as well as extensive rehearsals to transmit math concepts solely through the auditory channel. These are among the ideas that the students tested out. The basis of this curriculum is the ability to adopt numerous strategies without violating the fundamentals of mathematics instruction. The program's scope has grown significantly as it has inherited the spirit of taking on new challenges, exploring new possibilities, and utilizing existing funds to produce mathematics, making learning a pleasurable experience for students and teachers. Liberal management support and good use of technology are other crucial factors that might help such initiatives succeed (Ernest and Paul, 1989).

It is an innovative program commissioned in response to concerns about long-standing challenges in school mathematics education. The program is designed to prepare wholesome instructors who are both teachers and learners simultaneously. It enables teachers' wings to fly, investigate, and reach beyond the set confines of learning and teaching, where the sky is the limit (Rampal et al., 1998). However, the achievement of the program will be evaluated by the stakeholders in due time; whenever the course's products can make a visible improvement in the level of mathematics teaching, such programs will undoubtedly inspire many educators to speak their minds and assume inventions. It is now or never to rewrite the country's concept of educational mathematics:

> '*A classroom where no child is afraid of mathematics, where mathematics is a topic of discussion and cheers, where students choose mathematics by choice rather than by force or compulsion, where mathematics is an interactive session of discovery, dialogue, and learning without boundaries, and where mathematics learning is enjoyable for all.*

7.5 Need of Innovative practices in Mathematics Education

Mathematics is a discipline that is both practical and theoretical. Its goal in school is to create many abilities in students, such as critical thinking, solving problems, calculation, critical skills, logical thinking, logic (inductive and deductive), and so on. It also aims to build many life abilities of students. Consequently, the focus should be on rote learning (memorizing formulas, theorems, and the like) and utilizing their meta-cognitive capacity (ability to apply a formula or result in a specific problem area).

The curriculum at the school level, developed by approved bodies such as NCERT (for CBSE & other Boards) and various state board curricula, etc., is highly regarded. However, children's mathematics learning achievements continue to fall short of expectations. This demonstrates a lack of strategy development, new practices, and new learning and methodical planning for teachers for varied students. Teachers will need to devise strategies. Teachers should consider providing the study content in a sophisticated and unique style. Teachers can utilize new and creative ways to motivate pupils to study mathematics and, more significantly, to grasp it. It necessitates rethinking the nature of the mathematics curriculum, the quality of the techniques used to explain it, and the degree to which students understand their subject and earn excellent results after each semester.

In order to teach Mathematics, it is critical to use a variety of tactics and to be innovative. Though teachers discuss their restrictions or barriers to implementing such strategies in the classroom, including time constraints (due to the task of completing the entire syllabus within the allotted time), a low motivation, a loss of interest in learners, a lack of ongoing professional development on technology, or a lack of innovative teaching-learning reserves, among others, they must focus not only on completing the syllabus but also on practical ideas and inventive planning (Fisher, 2010).

7.6 Innovative methods in Mathematics Education in school

7.6.1 Inducto-Deductive Method

The inductive method involves progressing from clear instances to generalization, while the deductive method involves shifting from generalization to particular cases. Like most institutions, lessons start with abstract a concept which is beyond the grasp of the students. Equations, theorems, examples, and results are all produced, proven, and put into practice. Nevertheless, before continuing to generalizations and abstract ideas, the teacher must start with particular incidents and tangible things. The teacher must then explain how to create generalization and apply it to particular situations. This method helps students learn more effectively because they do not have to cram knowledge, which has a long-term effect.

Example: The sum of a triangle's two sides is more significant than its three sides.

(If you ask a pupil to estimate the edges of any triangle and add any two of them, the outcome would constantly be more significant than the third. After that, the educator can move on to the overall evidence).

7.6.2 Play-Way Method

This strategy incorporates number-related games and enjoyable activities.

Example: formation of Pascal"s triangle in solving n $(1+x)^n$

Students are unaware that they are studying, but their involvement in these actions results in them accumulating information in specific ways.

7.6.3 Laboratory Method

Only with the invention of electronics do most universities now have well-equipped desktop computers. The accessibility of computer processing software could be used to supplement classroom mathematical education to enhance students' energetic learner engagement; to swap long and arduous numerical and algebraic manipulations by communicating assisting rationale once attempting to answer math problems; to make innovative exercises manageable; to create problem-solving skills dealing with much more fascinating and complicated problems in terms of numerical, alternative, and algebraic manipulations; and to make innovative activities easier to manage. Teachers have a tough time creating and executing learning experiences that support these possibilities to be realized (Sidhu, 1995).

Computer programs such as Maple, Mathematics, Matlab, and Group algorithm program (GAP), durable software instruments intended to answer fundamental mathematical issues, could explain some math problems.

Problems in mathematics, physics, and engineering (among others) could be examined using these applications' built-in commands or by using these programming languages to develop one's personalized programs. They could be used to solve problems in Algebra, Calculus, Differential Equations, Statistics, Linear Programming, point charting in two and three dimensions, and creating a three-dimensional picture of an object, among other things.

7.6.4 Oral Presentation in Mathematics Learning

Because students' educational requirements vary, it is critical to examine the benefits of communication skills in mathematics service units. For kids with a "mathematical mind," accessing areas of the brain that traditional instructional methods ignore might be the answer. The multiple intelligences theory and brain-based education could be the resources these pupils need to acquire trust in their mathematical abilities. Oral presentations allow all students to showcase their abilities excitingly and uniquely. Curiosity kindled when researching the topic may lead to a renewed interest in arithmetic. With fewer students wanting to study mathematics, maybe an injection of creativity in service units will ignite the attention of these and other math learners."

Students and teachers had favorable remarks about the benefits and usefulness of introducing oral communication exercises into their everyday mathematics teaching. The oral presentation involves a vocal flow of ideas and understanding confirmation. To begin, this strategy is considered a different way for teachers to learn their students' mathematical understanding and make suitable instructional decisions. Second, it is viewed as a tool that learners could use to share more information. One of the main goals of oral presentations is for teachers to hear what students think about math and how they convey it, as well as their grasp of arithmetic in their own words. Moreover, teachers who use oral presenting activities must give students a chance to think through topics and difficulties, communicate their thoughts, show and clarify what they have learned, defend their viewpoints, and reflect on their understanding and other people's ideas.

7.7 Mathematics learning strategies for the 21st century learners

7.7.1 Blended Learning

Over the last few years, blended learning has gained acceptance in the educational profession. It is a formal education program in which a student learns, at least in part, through content delivery and instruction through digital and online media while maintaining some level of student autonomy. While it may seem like a significant jump for some teachers, there is a solid notion underlying the buzz that provides students some influence over their educational environment and can boost their motivation in math classes. Blended learning is not just scanning a paper into a computer or mobile device. It comprises arithmetic topic material that is important to students and allows them to self-pace their study in part. Instead of replacing teachers or direct instruction, which is still quite vital, technology enriches the learning experience and gives students at least some influence (Jacobson et al., 2012).

7.7.2 Mobile Learning

The generation of kids who grew comfortable with technology has a different attitude to learning than previous generations. They are accustomed to having instant admission to multiple sources of info, multitasking, and emotionally linked to peers via mobile devices. In the present notion of learning, the blend of mobile technology and the younger generation's "digitally enhanced" cognitive and social abilities will necessitate new solutions. M-learning is the use of ubiquitous handheld technology, as well as wireless and mobile phone networks, to assist, improve, and increase the range of learning and teaching. It has immense promise as a tool for promoting collaborative learning, engaging learners with information, as a substitute for books or computers, as an alternative to attending campus lectures, and for 'just-in-time' transmission of content in settings where learners are geographically scattered. It highlights that stakeholders have been highlighting the potential of digital technology for Mathematics for over two decades as having various possibilities. The introduction of mobile learning technologies into learning and teaching has provided teachers with both new opportunities and challenges." Mobile learning technology is 'naturally ubiquitous, wireless, easily transportable, and endowed with advanced features, giving a new dimension to curriculum delivery' (Jonassen, 1989).

7.7.3 Experiential Learning

Experiential learning is a teaching-learning approach in which students learn by doing and comprehending the subject's real-world applicability, allowing them to retain information for more time. By its very nature, this learning facilitates the growth of various skills, including planning, teamwork, coping with stressful conditions, responsibilities, and leadership.

7.7.4 Art Integration

Art-Integration was a cross-curricular pedagogical strategy that used many aspects and forms of art and culture as the foundation for facing conceptual learning across topics. If entrenched in classroom transactions, art-integrated education may create happy classrooms and assist art and culture in the teaching-learning procedure at all levels. Learners can demonstrate and apply what they have learned in a real-life situation by combining art and maths ideas. By March 8, 2019, CBSE has implemented Art-Integrated Learning in all its affiliated schools.

7.7.5 Interdisciplinary Approach

It is a method of curriculum integration that develops conceptual understanding via studying several subjects and their connections to the actual world. It refers to using activities to combine two or more disciplines to improve the clarity of a single concept.

7.7.6 Personalized Learning

Personalized learning is training that caters to the specific needs of each student through pedagogy, curriculum, and learning settings. The experience is personalized to distinct learners' diverse learners and specific interests (Strauss and Corbin, 1990). The learning objectives, material, approach, and pace, may all differ in a tailored educational environment. Personalization also includes outstanding education that encourages student advancement based on subject matter expertise. It is critical to know about every child's interests, limitations, and learning style after analyzing each student academically and knowing 'where they are in their learning development. To serve auditory, visual, and kinesthetic learners, the classroom might be divided into separate regions or stations based on their requirements and talents. Some stations, for instance, may enable inquiry-based, autonomous learning, whereas a separate area for group activities may be available. Material, skill, and evaluations can all be used to create groups.

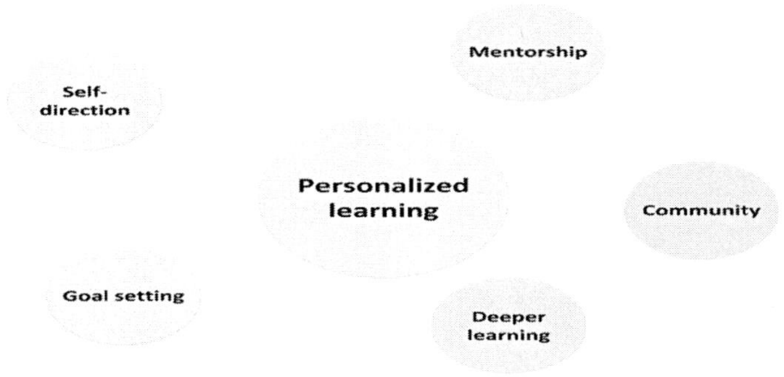

Figure.7.2 Personalized learning

7.8 Summary

Trends come and go, but a good education will last a lifetime. Various factors will be identified as problematic in mathematics instruction in our schools. We base our understanding of these concerns on the following four problems, which we consider to be the most pressing: Fear and failure in mathematics among some of the mass of youngsters; a curriculum that disappoints both the gifted minority and the non-participating majority at the same time; Valuation approaches that reinforce the impression of mathematics as a mechanistic calculation, and in the teaching of mathematics, there is a lack of teacher training and support. The dual principles that all students can learn mathematics and that all learners have to study mathematics underpin our vision of excellent mathematical education. So, to ensure good education for all, we must adopt creative teaching methods and learning tactics. As a result, we use cutting-edge teaching methods and learning tactics to provide our students with the best mathematics education available.

CHAPTER 8

COGNITIVE DEVELOPMENT

8.1 Introduction

The cornerstone for child education is knowledge of children's cognitive development. On the one side, current advances in mental growth cast doubt on some long-held beliefs about how children learn more effectively; on the other side, they give theoretical backing and empirical studies for several well-known ideologies in prior learning. The benefits of cognitive growth studies to child teaching were highlighted in essential aspects in this chapter. Understanding children's mental growth was the foundation of childhood education. On the one hand, recent discoveries in cognitive development doubt some long-held views about how children are taught and acquire more successfully; on the other hand, they provide theoretical basis and empirical data for numerous well-known basic learning philosophies. This chapter emphasizes some critical aspects of the advantages of cognitive development investigations to childhood education (Bukatku and Daehler, 1995).

8.2 The Individual versus the Environment

Whether variations in a person's knowledge could be credited to modifications or processes able to operate within the person themself, nativism, or whether such variations could be credited to environmental processes, empiricism, has been at the heart of all theories of development. This topic could be traced back to early disputes among ancient philosophers; Plato argues for nativism about notions like a triangle or God, claiming that there is no proof in the biosphere to judge these concepts.

On the other hand, Aristotle believed that extrinsic inputs inscribed all understanding on individuals. These discussions revived over the following 2,000 years with a new obvious form of empiricism, which was notably stated via John Locke's notion of a tabula rasa, in which he claimed, individuals arrive at the biosphere as "blank slates" completely molded by their surroundings.

On the other hand, the theorists Immanuel Kant and Jean-Jacques Rousseau concentrated on the native human character, with Rousseau claiming it was noble, pure, and biologically derived. To understand genuine human nature, one should seek individuals "outside social traditions." Kant contended that human brains have common characteristics and patterns that shape one's world perception. To properly appreciate human evolution, one must first understand how newborns and young children's internal, natural psychological structures interpret data about the world.

Current concepts of cognitive growth continue to wrestle with the fundamental concerns of the person against the environment; however, the focus has shifted from finding which impact is more essential to trying to justify how the two interact. Monitoring, research, continuous data capture, computational computer modeling to assess hypotheses via simulations, and collecting knowledge on the natural and neurological bases of cognition and behavior have become more evidence-based (Carey, 2009).

As questionnaires have improved, so have theories that are becoming more biologically plausible and capable of describing both how babies and children enter the world and how their cognition changes. The following sections outline the major hypotheses that have led to modern development methodologies.

8.3 Major theories

8.3.1 Domain-General versus Domain-Specific Theories

Within the current discussions regarding empiricism and nativism, two schools of thought have evolved. Domain-general theorems aim to discover generic cognitive growth tools that explain changes in all kinds of cognition inside a person, including "brain maturation" or "equilibration." Domain-specific concepts describe changes in only specific internal processes within a person, including numerical, understood, or linguistic. Modern empiricism is much further domain agnostic, whereas contemporary nativism is more domain particular.

8.3.2 Theory and Probabilistic Learning

Children's learning mechanisms are a topic that is profoundly entwined in the nature vs. nurture argument. When prompted by contextual stimuli, nativists consider learning as the emergence of innate, module-like structures, whereas empiricists reject the presence of abstract patterns and describe learning as the development of associations among context-dependent data. Developmental 'concept theorists,' based on Piaget's constructivism, have proposed that the method children study mirrors the understanding of technical theory construction (Gopnik and Wellman, 2012). By employing a probabilistic learning model based on Bayesian statistics, the theory aims to reconcile the nativists' and empiricists' perspectives on the process of students' education. The truth or falsity of statements is unclear in Bayesian statistics, and proof of the actual condition of the universe is represented as degrees of confidence. Theories are assertions that are made about causal structures. A hypothesis' last chance is given, and it is continually efficient in light of new information. According to the probabilistic study, even pre-schoolers may conclude causal arrangements from statistical input.

Real-world organizations could be characterized mathematically by multiplicative models, which enable the students to calculate sequences of data provided by the organization and draw conclusions. The students could also reverse the procedures and use evidence to study the organization of the actual world. The probability model was unique in that it assists the beginner in selecting the most likely assumption from between all the theories consistent with the probability patterns of proof. As a result, infants are not limited by innately defined illustrations, nor do they begin with a blank slate; instead, they constantly review their information and test theories like scientists do. Studying becomes a procedure of combining existing information and new information to revise experience-based world representations. Gopnik and Wellman (2012) presented empirical facts those students aged 16 months to 4 years employ probabilistic learning methods, including informal testing via playtime, mimicry, and informal pedagogy in their psychological and physical fundamental inferences in a recent analysis.

8.3.3 Theory of Mind

Children learn new causal links via first-hand inquiry and seeing and imitating others' interventions. According to Gopnik and Wellman (2012), the theory creation device of children's education applies not only to the temporal structure of the universe but also to the head of the reader instructing them. Indeed, learning and teaching require an intuitive grasp of one's and others' mental conditions and actions, including ideas, judgments, information, intentions, wants, and feelings, as well as the behavioral repercussions of these states of mind. Understanding people's mental lives and their association with behavioral outcomes was a vital developmental milestone throughout the early years of life for studying and being educated more broadly (Wellman and Lagattuta, 2004).

8.3.4 Testimony

Sobel (2010) stated that probabilistic education incorporates bottom-up and top-down processes, including knowledge extraction from observable data and direct teaching. Harris (2002) referred to top-down learning procedures as 'testimony learning.' Children rely heavily on the testimony of adults to learn, particularly regarding entities that are not visible, including metaphysical entities such as God and the tooth fairy, and scientific entities including the shape and germs of the Globe. The extent to which youngsters obligate to these entities is determined by the amount of information they are given. Sobel proposed that students learn fundamental structures merely by being taught about them. Nonetheless, children do not believe all they are expressed; instead, they preferentially believe the testimony of specific sources from a young age. Children regularly examine the informants' motive, dependability, competency, and smartness when learning by witnessing them at work; they also evaluate testimony with their perception and make rational decisions.

8.3.5. Conceptual Change

Theoretical change is another application of the concept to child teaching (Vosniadou et al., 2008). Scientific concepts emerge from paradigms in the scientific community that are webs of standard concepts, opinions, and practices. The evolution of scientific theory is revolutionary rather than incremental: old paradigms were discarded and substituted by new ones. The substantial differences among ideas make it impossible to conclude that one was superior to the other. Theory links the learning and development of children to the creation of theories in the scientific community. Domain-specific, hypothesis frameworks are used to gain knowledge. Significant changes in theoretical concepts contexts are involved in conceptual alteration. Previous information or misconceptions become stepping stones towards more precise science and technical clarifications throughout the conceptual shift process. However, they may coexist with correct concepts in interpretation and understanding. Individual comprehension varies between domains and contexts; there is not always a linear progression with stage.

8.3.6. Self-Regulation and Executive Function

Self-regulation was a general theory that created the base for growth throughout all fields, whereas the concept was a field-specific method of education. Self-regulation is a multifaceted concept that encompasses various psychological processes such as biochemical functions, concentration, emotion, behavior, and cognition. Self-regulation involves three parts as a goal-directed action over time and in many contexts: expressing objectives, inspiration to accomplish aims, and efforts to implement aims despite difficulties. The administrative purpose was a self-regulation notion mainly addressed by cognitive psychologists. It incorporates the cognitive characteristics of mental set changing, working memory, and inhibitory control. Whereas educator or parent bits of intelligence or behavior tasks including interruption of gratification could be used to assess self-regulation, cognitive processing metrics are primarily behavioral tasks for children, like cognitive tasks, Stroop-type jobs to assess inhibitory control, and set changing tasks like the Dimensional Change Card Sort (DCCS) task (Zelazo et al., 2003).

8.3.7 Domain General: Cognitive Development Behaviorism Foundational Psychological Studies

In the early nineteenth era and early parts of the twentieth era, psychological research contributed data determining the method to the terrain of cognitive growth concept. Due to a reluctance to infer or form hypotheses about internal mental procedures, behaviorism was favored during this time frame of a psychology experiment; a theory predicated on the idea that exploratory or empiric studies cannot provide direct information on mental procedures, so one should solely focus on detailing behaviors in contextual factors. B. F. Skinner and other behaviorists were empiricists, stating that growth could be described as acquiring new knowledge via classical and operant conditioning. Through these processes, humans and animals build links between environmental stimuli and their behavioral and other responses. Despite its continued influence, the growth as a knowledge concept was restricted in its skills to describe or adequately characterize kids' cognitive development paths.

8.3.8 Vygotsky's Sociocultural Theory of Development

Lev Vygotsky (1896–1934), who drew significant emphasis on the function of caregivers and culture in the developing atmosphere, agreed that environmental and biological effects interact. Vygotsky's theory emphasized the importance of social contact in children's cognitive development, particularly engagement with expert social partners.

8.3.9 Dynamic Systems Theory

8.3.9.1 Contemporary Development Theories those are Domain-Specific

According to field-specific concepts, cognitive growth occurs differently in various fields. That idea appears to be in line with current editions of constructivism, which asserts that, because growth happens via connections with nature, growth can occur at different rates in various domains, relying on the value and amount of a kid's conversations within that field. Recent nativists, on the other hand, have claimed that growth starts with intrinsic or biologically determined predispositions to study in particular ways in precise areas, laying the groundwork for learning and broader growth based on real-world experiences.

8.3.10 Contemporary Nativism: Modular Foundations

Jerry Fodor and Noam Chomsky are the two primary theoretical foundations of contemporary nativism. Chomsky examined the basics of language progress and proposed that behaviorist models of education by relationship might not be able to account for the creation of complex linguistic organizations. The poverty of stimulation (POS) argument asserts that there are insufficient stimuli. Jerry Fodor and Noam Chomsky are the two primary theoretical foundations of contemporary nativism. Chomsky examined the basics of language acquisition and proposed that behaviorist models of learning by connection might not be able to account for the creation of complex language structures. The poverty of stimulation (POS) argument asserts that there are insufficient stimuli.

8.3.10.1 Contemporary Domain-Specific Theory: Core Knowledge Theory

According to the core information concept, a set of natural abilities, or learning strategies, permits a quick learning experience in some domains but not others. Core cognitive limitations that structure a kid's thinking to promote information acquisition in at least four fields have also been postulated by contemporary developmental scientists: objects (naive physics), numbers, space (naive reorientation or geometry), and people (naive psychology).

8.4. How Do Children Learn?

So far, the chapter has concentrated on learning appliances that develop via time. What about the actual education process? What was the best method for youngsters to learn? As per Siegler (2000), with the demise of associationism, study on students' education has reemerged; focused on the education of expressive ideas and abilities which are significant in children's existence. The overlapped waves hypothesis (Siegler, 2000), which focuses on the range of tactics and ways of thinking in learning and problem solving, is one of the most popular ideas on students' education. Numerous strategies coexist simultaneously, according to this theory, with new tactics being introduced and old approaches occasionally ceasing to work, and a specific approach could remain for a long time, even after more advanced approaches develop. That stresses cognitive variability in learning, both within and across people, even when a specific is addressing a similar issue in several trials or a similar test.

Microgenetic learning approaches are used in the overlapping wave's theory to analyze the intricate learning experience with ecological validity. These techniques make much information about the rate of alteration throughout a time of quickly changing competence and then analyze them thoroughly. The overlapping wave's theory can examine learning in 5 dimensions: path, rate, breadth, source, and variability, thanks to the abundance of micro genetic data.

The overlapping wave's hypothesis connects learning and development, one of its significant achievements. Training and development processes, according to Siegler, have a great deal in mutual in terms of diversity, optimal, and alteration, and hence cannot be differentiated from one another. Sure the learning studies resemble the age-linked deviations identified by developmental psychologists such as Piaget. Siegler refers to his theory as a progressive learning method that considers both quantitative and qualitative shifts. Microgenetic study has educational implications since it indicates how coaching instruction affects children's learning and thus aids in developing more efficient development instructions.

8.5 The Brain and Cognitive Development

You have probably read about the brain and nervous system in an introductory psychology class. You are probably aware that the brain is divided into various sections, each responsible for specific processes. The feathery-looking cerebellum, for example, regulates and orchestrates balance and smooth, skillful movements—from a dancer's exquisite gestures to the ordinary achievement of consumption without intense yourself in the nose with a fork. Higher cognitive functions, including learning, might be influenced by the cerebellum. The hippocampus is responsible for remembering new knowledge and recent events, whereas the amygdala is in charge of emotion control. The thalamus influences our capacity to learn new, mainly linguistic information (Johnson et al., 2005).

Scientists now have unprecedented access to the functioning brain because of advances in brain imaging technology. For example, when infants or adults perform various cognitive tasks, functional magnetic resonance imaging (fMRI) displays how the blood circulates in the brain. If people undertake activities like reading or acquiring vocabulary words, event-related potential (ERP) measurements are taken through the brain's skull or scalp to monitor electrical activity. PET scans can be used to track brain conduct in a variety of situations. Let us start by looking at the brain's most minor parts: neurons, synapses, and glial cells (Fischer and Kurt, 1987).

8.5.1 How the Brain Works

What do you think of when you think about the brain? Is the brain a culturally agnostic vessel that stores information in a similar system for everybody? Is the brain similar to a collection of evidence or a machine that stores data? Do you have a morning routine? Transfer everything you require for the day, and then do your business. Is the mind similar to a channel that transfers knowledge from one place to another, such as an instructor to a student? Kurt Fischer, a Harvard professor, and evolving psychologist, has a distinct perspective created on neuroscience studies.

Play and purposeful practice, formal and informal learning, all shape the brain (Dubinsky et al., 2013). Plasticity is a concept that represents the brain's ability to modify neurons, synapses, and activity continuously. Cultural changes in cognitive movement are instances of how the brain is shaped by encounters in the world via plasticity. In one research, Chinese speakers who added and evaluated Arabic data suggested activity in the motor parts of their brains, but English talkers completing the identical responsibilities exhibited movement in their linguistic skills (Tang et al., 2006). One possibility was that Chinese students were educated in arithmetic by an abacus, which is a computation device that requires motion and spatial placements. Those children preserve a visual-motor sense of numbers as adults (Varma et al., 2008). In terms of how languages influence reading, there are also cultural variations. Native Chinese talkers, for instance, stimulate supplementary portions of their brain related to spatial evidence interpretation when they read, which is likely because the linguistic characters employed in inscribed Chinese were images. However, when Chinese talkers read English, these spatial parts of the brain are also activated, suggesting that reading skills can be achieved through various neurological routes (Hinton et al., 2008).

8.5.2 Neuroscience, Teaching, and Learning

Let us start by defining what instructors do not know about neuroscience. Numerous general neuromyths exist. We must be cautious about what we see in the mass media. This was not a myth that teaching could alter the brain's structure and organization. Deaf persons who use sign language, for example, have unique designs of electrical activity in their brains than deaf people who do not use sign language (Varma et al., 2008). What other consequences does instruction have on the brain?

8.5.3 The Brain and Learning to Read

As learners, learn vocabulary, brain imaging study shows interesting variations between proficient and less-talented readers. For instance, according to one imaging study, ERP assessments of electrical activity in the brain revealed that low-skilled users had problems generating maximum-quality images of new language terms in their brains. Less-skilled readers' brains often did not recognize the new term when they met it later, even though they had learned the phrases in a previous lesson. We can see how it would be difficult to grasp what you read if the terms you learned afterward look unfamiliar. Bennett Shaywitz and his coworkers (2004) studied 28 excellent learners aged 6 to 9 and 49 children aged 6 to 9 who were wrong readers in some other study. MRIs revealed changes in brain movement between the two groups. The poor learners used areas of their left brain hemisphere inefficiently but sometimes overloaded their right hemispheres. Reading skills increased after almost a hundred hours of intensive teaching in letter-sound mixtures; wrong readers' brains began to behave more like those of intense viewers and remained to do so a year later.

Slow readers who underwent regular school remediation did not demonstrate any improvements in brain function.

8.6 Piaget's Theory of Cognitive Development

Jean Piaget, a Swiss psychologist, was a child prodigy. He produced numerous scholarly papers on mollusks (marine animals including clams, oysters, snails, octopuses, and squid) while still in his teens that he was given a job as the director of the mollusk exhibit at the Museum of Natural History, Geneva. Also, he stated to museum executives that he needed to end high school first. In Paris, Piaget was employed in Alfred Binet's laboratories for a time, producing children's intelligence tests. The justifications youngsters provided for their incorrect replies piqued his interest, prompting him to research the reasoning behind their responses. This question piqued his interest for the rest of his life (Ojose and Bobby, 2008).

Piaget developed a model detailing how people gather and organize information to make sense of their reality throughout his long career (Piaget and jean, 2000). We will look at Piaget's concepts in-depth since they explain how children's minds evolve from infancy to adulthood.

Specific forms of reasoning which are easy for adults, including the Pittsburgh question in Stop and Consider, were not so easy for children, according to Piaget. Do you recall the 9-year-old boy asked if he could be a Genevan at the start of the story? "No, that is not possible," he replied. I am a Swiss citizen presently. Around the same time, I cannot be a Genevan." Consider instructing this student in geography. The learner has difficulty categorizing one notion as a subset of another (Switzerland). Another distinction exists between adult and kid thinking. Students' understanding of time might vary from yours. They might believe, for instance, that they would one day catch up to a sibling on stage or that the past and the future are interchangeable. Let us have a look at why that is.

8.6.1 Stages of Cognitive Development

All children, as per Piaget, went through four similar phases in the same order. According to Bukatku and Daehle (1995), there was a different level of thinking, internal organization, and interpretation of geographical data and events at every stage of cognitive growth. Piaget's theory clearly illustrates that a children's knowledge is solely essential for the performance that she or he has reached, and educators should retain this attention when teaching children at several ages of intellectual development.

8.6.2 Sensorimotor Stage birth to 2 years (infancy)

Is the first period of a child's learning and growth? Children require a fundamental structure for organizing and reacting to their environments, and their performance is circular. They also have a basic comprehension of what was successful around them. That is during this phase when children study to speak, which helps them develop socially and intellectually. These steps mention the children's level of thought or intelligence as evidenced by her or their activities. A child's schema was simple, limited to what the infant could determine with his or her sense organs. The steps last from when a child is born until he or she is roughly two years old. Cherry, Kendra (2014) as a child grows, it has an innate desire to arrange its environment. Even though the foundations take shape as a consequence of secondary circular reactions, some techniques of classification and experiences take shape due to secondary circular reactions. At this stage, the infant begins to appreciate the permanency of hidden or visible items. This is also the stage when youngsters exclusively see the world through their own eyes. When a youngster interacts physically with his or her environment, he or she creates reality and learns how it works (Lazarus, 2010).

8.6.3 Preoperational stage two to about seven years

At this stage, a youngster could think and give a logical line of belief. The child uses objects and symbolism to reflect something which exists in a concrete form, including a toy car that is interacted with as if it were a racing car. The youngster could not understand abstractly and wanted actual physical situations. It is also the development of semiotic processes that assist in language production. Because the child could work with symbolism, the children's communication, imagination, thinking, and solving problems grow at a higher rate during this stage. Even if the object's attributes were changed and seem distinct, the youngster may recognize them. The kid found it difficult not to accept the facts in front of their sight at this stage. Children's language expands, and their statements go from one and two-word sentences to entire phrases. Children could reflect on several viewpoints and consider numerous perspectives. Animism, transudative reasoning, egocentrism, syncretism, lack of categorization, lack of seriation, decentering, conservation abilities, and quick learning of language were all features of the pre-operational age child (Lazarus, 2010).

8.6.4 Concrete operational stage: from seven to eleven years

A child can apply logic and reasoning procedures on the foundation of credible proof. In formal operations, children were said to think in ideas of concepts and abstractions, and concrete facts. Resolving problems and thinking were adequate to live the remainder of one's lifetime at this age. A child was suitable for creating conceptual frameworks to describe bodily involvements and solve abstract problems. Arithmetic equations, for instance, could be handled with numbers rather than things. At this age, the youngster was intelligent, reasoning depending on past experiences and concrete suggestions. Throughout this age, the kid could effectively complete tasks relating to matter conserving, transitive thinking, and sorting techniques (Lazarus, 2010).

8.6.5 Formal operation stage: from eleven years upwards

Not only was reasoning abstract, but that was also rational. The presence of concrete items was not required for the cognitive to take place. A student can now create potential answers to issues in a methodical method. In that stage, the societal context was increasingly crucial. Concrete instances are essential to aid children's comprehension of abstract relations. The stage happens in initial adolescence, and the youngster is more abstract and shallow thinking at this time. The student's reasoning structures were similar to those of a mature at this point and contained conceptual intelligence. That was the child's most significant level of reasoning, and he or she was skilled at thinking beyond the obvious. Also, in this stage, the student could pay attention to their care for things that do not exist. The youngster can now complete a broad range of responsibilities requiring hypotheses. The learner's opinions can be refined by putting the student in a scenario where they must solve difficulties (Lazarus, 2010).

8.7 Educational implications

Cognitive growth cannot be explained solely by active participation and exploration—inheritance and maturation, and also experiences. Learners' cognitive growth is built on active participation in investigating their physical and cultural worlds. When the instructor optimizes things, learning and teaching must be dynamic, experimental procedures in which students question, discuss, reflect, and resolve complications for them. Uneven cognitive development and reasoning growth was an unequal procedure in which individual learners alteration gears at various times learning parts, and the requirements of the social setting in which the kid is growing affect the learner's growth. The teacher interprets the Piagetian stage as a restriction rather than a developmental potential. Teaching focused on potential is considerably diverse from teaching with attention to boundaries. The distinction in thinking quality among children and adults has significant ramifications for moral, social, and sentimental areas of development. These features relate to a child's definition of social relationships and how they deal with social, moral, and emotional problems as they grow (Lazarus, 2010).

8.8 Summary

The process of learning is inextricably linked to the growth of children. Learning is a probabilistic model-based new theory testing procedure. Children are learning, monitoring, and explicit instruction, during which time they are constantly exposed to spontaneous experimenting, assessment of other people's intervention, information, and purposeful attribution in teaching. Self-regulation and executive function are domain-general cognitive talents that help students achieve academic success by helping them create learning aims, turn their efforts to the educational activities at hand, resist temptations, and regulate their feelings. Piaget holds a constructivist perspective, believing learners actively participate in their knowledge. As per Piaget's thesis, A syllabus that supports intellectual development by providing explanations and logical stages is required for students. He also indicated that students could only acquire particular subjects during particular stages of cognitive development. Piaget highlighted the significance of learners' actively engaging in learning, so instructors must observe students actively participating in what has been taught in class.

CHAPTER 9

COGNITIVE NEUROSCIENCE IN MATHEMATICS

9.1 Introduction

The popularity of neuroscience research has prompted numerous individuals to look to the brain to seek solutions to learning problems. Cognitive neuroscience, an integrative field of study focusing on human thinking, can contribute closer to the brain and learning. The biological brain is the center of neuroscience research. However, understanding education and knowledge through examining the anatomy and functioning of the brain may not be quite as simple as other cross-disciplinary interactions. When neuro-scientific approaches were first applied in cognitive processing, educators were skeptical of how much could be learned by learning the brain. "Neuroplasticity," which indicates that the brain could be modified during birth or shortly after experience or environments, is a critical concept in cognitive neuroscience that is especially relevant to cognitive theory. The positive element of neuroplasticity is that it indicates the brain can grow, but it should not be ignored that it can be for good or the worst (H. W. Lee and Juan 2013).

Learning in cognitive neuroscience explores the neural system that underpins multiple cognitive activities. Neuroscience, according to research, may also aid in the advancement of education by offering more information on the neurological basis to facilitate development and refine cognition theoretical approaches. There are two main points cognitive neuro-scientific study could enhance: To begin, it might explore the underlying brain circuitry of cognitive abilities at the basis of subcomponent capabilities, based on current cognition models produced by cognitive psychologists. Secondly, it can go further in developing brain-area-to-cognitive-function mappings to look into how activity patterns are linked to education.

9.1.1 Cognitive Neuroscience in Education

In cognitive psychology, there is a differentiation between verbal and nonverbal memory functions and visual and auditory cognitive ability within both categories (De Smedt et al. 2010). By using functioning brain imaging, a great deal of study has been done into how parts of memory recall, auditory processing, and the application of techniques and the structure of cognitive performance are linked to cognitive function (fMRI or PET). The design and evaluation of modern learning environments, characterized by a concentration on (structured) self-directed studying, and the acquiring of perceptive, intellectual, and cooperative learning, has been a prominent focus of academic research. Some of the most common teaching topics and challenges include:

- Multimodal computation effects have to learn from multiple representations, cognitive effort, the involvement of knowledge in problem-solving, implicit and explicit learning, self-directed knowledge, regulatory skills, and thinking skills, and the use of observation and representation for studying to accomplish practical and cognitive tasks.
- Motivational mechanisms and their significance in education.
- Learning issues, such as dyslexia and dyscalculia.
- Studying specialized subjects, such as (second) language learning and mathematics.

9.1.2 Cognitive Neuroscience in the Classroom

There are several analogies to education, memory, motivation, cognitive development, and other topics, but none to education, learning, children as students, or pedagogy. Because cognitive neuroscience enhances the knowledge of learning principles, educators must be aware of the implications and applicability of education in formalized educational contexts, particularly in school classrooms. A return to the principles of teaching and learning could even help retrieve the learning goal from politicians and corporate executives whose primarily instrumental goals for schooling and higher education have produced so much discontent among teachers in recent years. To put it another way, one positive motivation for educators to pursue cognitive neuroscience is the expectation that it will help instructors become less sidelined as educationists. Cognitive neuroscience information should be part of the knowledge base that enables such re-empowerment (Geake and Cooper 2003).

Given contemporary worldwide political and financial demands, especially from the information and communication technology (ICT) industry, replacing human education with online information extraction may be necessary. Teaching will continue to be primarily a human institution, and educators are always interested in learning more about the various elements that influence their students' learning. According to the researchers, such teaching should include an awareness of cognitive neuroscience achievements. As a result, education should implement an interactive bio-psycho-social paradigm that can only be achieved if educators and cognitive neuroscientists interact in communication to share their professional knowledge.

9.2 Cognitive Neuroscience Meets Mathematics Education

Mathematics learning is a vast subject of study. At the same time, the area would have its own questions that permit other areas of study to contribute to the solutions. Because the topic is essential in mathematics learning and instruction, maths itself – with its unique nature, structure, history, and social position – is essential for the study. The significance of maths concept for mathematics education, even though it can be an individualized philosophy: "In truth, whether one wants it or not, all numerical education, even if irrational, depends on a field of mathematics." The old theoretical analyses can reveal the state of graphical representations in mathematics (Schlöglmann 2008).

On the other hand, learning cannot be perceived only through the views of the individual student. Because learning occurs within social and contextual factors, discoveries from cognitive psychology are essential to comprehending the learning experience. Moreover, learners take place in a specific institution (a school), and school learning is always combined with teaching. As a result, pedagogical advancements are crucial to mathematics education development.

In cognitive neuroscience, there are two types of memory: declarative or explicit memory and non-declarative or implicit memory. Declarative memory is associated with consciousness. Therefore its information may be stated, whereas non-declarative memory's information cannot usually be reported. Moreover, the engagement of various brain centers is required for the two kinds of learning. Both are separated into memory subtypes: declarative memory is made up of episodic and semantic memories, while non-declarative memory is made up of memories for abilities, behaviors, prompting, habituation, and non-associative learning.

On the other hand, emotional learning is implicit learning based on mechanisms in the brain. It is necessary to remember the neuro-scientific reality that just because an individual cannot learn using declarative memory does not mean they cannot learn with non-declarative memory. Researchers can identify sub-processes required for the ultimate effectiveness of cognitive abilities, such as interacting with others, thinking about a problem, cognitively trying to calculate something, or solving a problem. A few issues in mathematics are sketched, and neurology may be able to elaborate on them.

9.2.1 Dyscalculia

Everyone possesses an inherent skill known as "number sense." That enables children to examine the physical factors in quantifiable form, allowing them to differentiate small numbers and recognize changes in a limited number of things. The number sense is an approximation mechanism; its results are usually never sharp, but it resembles the pointed answer. The number sense appears to be aided by certain parts of the inferior parietal cortex. Humans took a significant stride forward with the establishment of representational number systems. These technologies enable us to construct "cultural mathematics" by overcoming the limits of our basic number sense. Students should learn numerical terms and use them to quantify objects and perform calculations. However, some individuals have developmental dyscalculia, an intractable mathematical disadvantage. Dyscalculia may be caused by an irregularity in the right parietal cortex, according to neuro-scientific research. Such findings could aid in distinguishing between distinct types of dyscalculia, such as physiological and environmental dyscalculia.

9.2.2 Learning without Understanding

Many claimed that they consistently obtained poor maths grades and had little understanding of the subject. Educators and students frequently believed that the best way to study math and pass exams was to memorize principles and procedures and practice implementing them to mathematical problems. When educating slow learners, memorization is a typical method, but it undermines the belief that learning – – particularly maths – involves knowledge. Teachers also noted that the knowledge students gained via such a learning method was frequently rigid. For example, the student cannot complete the exercise even if the variables' signs are altered. A neuroscientific notion commonly employed to describe the phenomenon of prompting – mainly, the idea of a perceptual representation system – could focus attention on the impacts of rote learning (PRS). The PRS lets us recognize familiar objects and phrases on printouts (note that a PRS exists for each of the other senses, too). The PRS is designed to handle the shape and structure of language and entities while "knowing" nothing about their meanings or applications. Moreover, one can deduce from the PRS's features that the recovery process relies highly on perceptual aspects. One can understand why modifying the characters used to represent variables, or rearranging the word sequence of word problems, contributes to an inability to solve the "new" work if pupils use their PRS to manage mathematics problems.

9.2.3 Errors in Affective Situations

Long-term mechanisms and cognitive and affective characterizations that establish the evolution of the educational process are the focus of mathematics education study on the role of impact in mathematics; however, researchers need to have more insight into the impact of emotions during solving problems to clarify the occurrence of mistakes, particularly easy slip-ups. Cognitive ability is a concept that can assist us in comprehending these activities. As previously stated, the capability of all active memory elements – the storage systems, the episodic buffers, and the executive function – is limited; however, it is essential to remember that focus is necessary for all cognitive operations, including cognitive functions. Because researchers are aware of the sentiments in emotional situations, these sensations manifest in cognitive function and utilize memory space. One can suppose that because mental processing capacity is restricted, there may not be enough space for the facts needed to support the thinking activities. During the cognition solution process, attention focused on emotions rather than the cognitive and heuristic system elements. It is logical that the number of mistakes, especially slip-ups, would rise in such a setting.

9.2.4 Affective Categories

The subjective classifications of "emotions," "attitudes," "beliefs," and "values/ethics/morals" have emerged as a result of research and can be used to understand a complex situation. Quantitative research approaches disclose steady and less severe categories, whereas qualitative research methods can acquire fast-changing and extremely intense responses. Despite this, the research methodologies cannot distinguish between two categories. As a result, there are no widely accepted classifications for each. Furthermore, while the traits "stability" and "intensity" help to describe the groups, they are insufficient to solve the difficulty of differentiating them. Neuroscience differs among several brain systems that perform distinct tasks. There are separate systems for cognitive and emotional activities. These systems are found in many areas of the brain, and each has its memory system with unique characteristics. The physical and emotional systems are strongly intertwined, and an active emotional system can result in physiological reactions.

Moreover, even though the emotional system operates invisibly, one can detect a few of the end outcomes of an active emotion system, such as what we refer to as "feeling." Although when we can perceive emotional reactions in conversations, the study methodologies utilized to examine the categories of "beliefs," "attitudes," and "values" are dependent on cognitive recall; therefore, the techniques are controlled by cognition. When they employ qualitative approaches to study problem-solving circumstances, however, one can detect the impact of the emotional system and actions that are not directly controlled by cognition.

9.3 Developmental Cognitive Neuroscience of Arithmetic

Mathematical ability is perhaps one of the most necessary cognitive capabilities for a student to learn. What changes require some time as youngsters gain more complicated numerical thinking skills? What factors contribute to children's marked individual disparities in mathematics skills, or why do they exist? For years, developmental and educational researchers have been pondering such problems. Researchers are now in a rare position to address those concerns due to better quantitative neuroimaging and focused cognitive science (Menon 2010). Mathematical skills are built on a basic arithmetic knowledge system, which is usually in existence by age five and allows for such representation of numerical quantities using mathematical concepts.

Different aspects of mathematical cognition have been investigated, involving (1) retrieving, (2) calculation, (3) reasoning and decision-making concerning mathematical relations, and (4) solving intervention among several contending answers (interference resolution). It clarifies whether areas of the brain are constantly and significantly involved throughout mathematical activities, which sections give arithmetic assistance, and which areas of the brain assist arithmetic learning.

9.3.1 Implications for Learning and Academic Achievement

The growth of numerous fundamental processes is required for students to learn mathematics. A few of these functions, particularly decision-making and working memory, take a long time to acquire. Although the extent to which two kinds of knowledge contribute to steady differences in brain reaction is unknown, relieving the prefrontal cortex of computation complexity and thus making helpful processing resources accessible for more complicated problem-solving and reasoning is an effective way to promote learning abilities and skill acquisition. Education tactics that emphasize repetitive achievement, resulting in more automated recovery, may help develop a knowledge base while reducing the cognitive burden of the prefrontal cortex. By changing the interface form and intensity of calculations, cognitive neuroscience researchers could separate cognitive and neural mechanisms regarding mental math. Cognitive neuroscience has also highlighted the relevance of cognitive function, episodic and semantic memory, decision-making, and activities focusing on precise task completion and enabling the development and growth of arithmetic abilities.

9.4 Cognitive Neuroscience Contribution in Mathematics Education

Information on brain function may combine towards a more lengthy explanation of the various cognitive (sub)processes that occur throughout mathematical development and reasoning; that is one of the ways that cognitive neuroscience can contribute to maths education investigation that is outlined by the achievements in the special section. Neuro-scientific knowledge may supplement or expand the knowledge gained only from data acquired, for instance, by methodological redundancy (Smedt and Verschaffel 2010). During the identification of small and large numbers, various mental abilities take place. Individual variances in these math-related cognitive abilities should be noted, and cognitive neuroscience may help to comprehend these dissimilarities in mathematics achievement.

The latter measures enable the online and continual recording of cognitive allocating resources (i.e., information storage and processing capacities) throughout problem-solving, which is difficult to identify using accurate and response time data. Cognitive neuroscience may help to comprehend the complete range of individual variations in mathematical learning (from weak to proficient) by pointing to potentially (neuro) cognitive elements linked to these variations. However, most evidence indicating a link between brain anatomy or functioning and personality factors in arithmetic is still cross-sectional. Based on such data, it is difficult to say if individual factors in the structural and functional brain are the cause or the result of particular variations in mathematical skills. The neuro-scientific techniques, methodologies, and ideas have the potential to link and expand the information on cognitive processes, which can be gathered by using only behavioral approaches, which is a frequent strategy in academic research.

9.5 Cognitive Neuroscience and Teaching

The neuroscience research on mathematics highlights the connection between practicing arithmetic versus teaching or learning maths. Most of the existing research in neuroscience has centered on solving math problems. Consider the instruction of mental arithmetic; mental number lines are one of the most common activities used by educators. Consider asking students how many candies are left after three are taken away from a total of five (K. Lee and Ng 2011). When the numbers are tiny, such a question is answered by counting or straight retrieval. When higher numbers are involved and algorithms must be utilized, such as subtracting three sweets from 500, it becomes much more difficult. One method, particularly for slow learners, is to have them visualize a number line and count backward from 500. It is first offered in lesser quantities to utilize such mental number lines. The learners then continue to build and expand the number line. Such exercises frequently have pedagogical benefits. Is this ease of access due to the development of pupils' number lines, the fact that such tasks are entertaining and engaging for children, or some other factor? These are critical questions because they will assist in establishing the practical bounds of various pedagogies.

9.5.1 The Neuroscience of Pedagogy

It is essential to determine the neurocognitive basis of particular numerical computations, but it is also essential to investigate how these processes are learned. Formerly, educators were primarily responsible for transferring research evidence to the classroom. In instructional research, cognitive neuroscience can play an important role. The neural implications of learning-by-strategy with learning-by-rote, for instance, subjects were instructed in one of two methods over a week. The findings revealed that learning-by-rote affected the left angular gyrus, presumably reflecting the method's dependence on language. The precuneus was active during learning-by-strategy, which the researchers related to the employment of visual imagery.

Primary students (9 to 11 years old) are instructed to answer arithmetic word problems using the model method, which is introduced in many places using symbols or formalized algebra. Understanding such algorithms helps students find the proper solution to specific mathematical comprehension questions without continuing to interact with symbolic algebra's graphical and transformative tasks. Some learners choose a combined analysis method, which combines algorithms with features of symbolic mathematics. Others create a model picture and the algebraic equation representing it before resorting to mathematical techniques to undo the process. Students with lower inhibitory skills had difficulty applying symbolic mathematics. Learners who used the model method were also more willing to use it again. Some senior educators' views on the model technique reflect the possible difficulties of implementing pre-algebraic approaches. Many educators were dismissive of the system, believing it to be a barrier to children learning symbolic algebra. According to one of their worries, the model method is immature and non-algebraic.

9.5.2 Pedagogical Implications

If symbolic mathematics is more demanding on cognitive resources, one pedagogical consequence is that it is desirable to learn the system will help in elementary school and wait till children are much more cognitively developed to introduce symbolic mathematics. It is important to remember that the individuals in neuroscience were all individuals who were equally skilled in both techniques when evaluating. Similar variations will be observed in younger students; therefore, adopting symbolic algebra may necessitate even more work for early algebra students. Nonetheless, these are actual theories that need to be investigated more. Educators are concerned about students' use of mixed methods because symbolic maths is required to address issues in advanced maths and science. Learners who use mixed methods are prompted by the program to generate comparable algebraic expressions, starting with defining the factors given an algebraic word problem. Immediate feedback is provided by the built-in self-checking mechanism, which aids in the development of algebraic equations.

Concentrating on the educators, an assessment of these sessions revealed that they delivered the knowledge using a distribution model. Learners were taught how to convert variables instead of the processes for developing a solution of similar linear functions. Teachers reported that their approach was restricted by the short curricular time given to the instruction of symbols manipulations and transformative actions in post-lesson assessments. One specific goal is to implement solutions that will support learners in reducing the requirements of symbols mathematics on their working memory.

9.6 Cognitive Neuroscience and Mathematical Reasoning

The mechanisms involved in mathematics problem solving and reasoning have been associated with cognitive neuroscience research, which has concentrated on mathematical reasoning. Researchers looked at different elements of mathematics computation, including processing, computation, reasoning, and decision-making regarding mathematics relations, that "assisted in specifying which brain areas are analytically and continuously involved during mathematics tasks, which fields provide a supporting role in arithmetic, and which areas of the brain contribute to mathematics learning. Much of the cognitive neuroscience study is based on psychological research that reveals a few cognitive processes in mathematics learning (Stewart 2017). The research involved a school-based investigation with 90 students in the 4th and 8th grades to examine the possibility of using the Near-infrared Spectroscopy (NIRS) technique with students. Students sat beside a computer monitor while wearing sensor sets on their foreheads. Behavioral data included failure rates and response time, while NIRS focused on oxygen and blood levels in specific subregions.

Interestingly, both age groups showed identical functional brain rates. The researchers hope to see more initiation of brain regions associated with processing for older students, given that they were theorized to use retrieval strategies significantly and because they were pretty developed in solving word problems due to more excellent knowledge. These findings led individuals to conclude that: "even as significant cognitive arithmetic [involving two-digit addition in arithmetic problems], could establish from elementary to high school in aspects of faster calculation," The research that was the first to evaluate such significant numbers of students in such a short period (14 days) in an academic environment, also demonstrated the utility of NIRS technologies for investigating young children's simple mathematical activity.

9.7 Numeracy and Mathematical Learning Cognitive Neuroscience

Numeracy, like reading, is a suitable subject for cognitive neuroscience and pedagogical research since that is the product of the interaction of biology and experience. However, no widely accepted definition exists; mathematics entails an idea of the basics of numbers and numerically reasoning skills. As a result, it is the foundation of both elementary and complicated maths (de Jong et al. 2009). It is vital first to get a deep knowledge about those cognitive capacities to create a curriculum that assists students in developing their fundamental cognitive skills. Depending on the conclusion that even children have a certain number of skills, some study has suggested that mathematical abilities, like linguistic talents, are natural to humans.

Although research findings from the 1980s and 1990s concluded that infants could distinguish between two and three dots and operate basic arithmetic operations like 1 + 1, there were critiques that such experiments did not accurately regulate continuous variables correlated with numerosity. Though infants have the numerical ability, more carefully controlled research outcomes suggest that these skills appear limited to big mathematical measurement discriminating, for example, 8 and 16 sounds or dots.

There are two possible cognitive mechanisms to analyze numerosity: one for accurate memory of small numbers of items and another for approximated computational complexity. The numerical analogy phase, often known as the dual system, is considered active when the symbolic application of mathematical processes involves Numerical digits or numeral phrases. Many countries are moving away from actual teaching, mainly centered on deliberate practice, and toward more practical mathematics teaching focused on constructionist ideas in recent years. On the other hand, direct instruction has been shown to improve results for children with intellectual disabilities. Furthermore, because realistic mathematics education emphasizes skills that students with mathematical learning disabilities may struggle with, such as vocabulary, reading level, and math fact fluency, there is some debate about whether this method is appropriate for these students.

9.7.1 Training of Numerals in Neuroscience

Understanding the neurological basis of numerical cognition can help understand better how training and teaching affect performance. The activity of inferior frontal areas decreased after training with both multiplication and subtracting, implying that training lessens the demand on working memory and executive control. Multiplication training also increased activity in the left angular gyrus, indicating that the method shifted from calculating to more spontaneous retrieving. A better understanding of the evolution of math skills from the area of cognitive neuroscience may make it easier to create research paradigms in academic research. Some kids with mathematics learning challenges, for example, appear to adopt immature and ineffective methods. Such methods can be ineffective since they put more strain on cognitive processes like working memory.

However, strategy use is frequently established by children's written or verbal assessments, and there is reason to assume that such assessments may not accurately represent strategy use. Brain imaging might be utilized as an objective measure that, when supplemented with much more qualitative information, could provide indications of method application, allowing for more profound research into strategy efficiency and the benefits of strategy training. Such a study could address whether practical vs. direct mathematics instruction is more successful, especially for children with low IQ and specific cognitive disabilities. Brain imaging methods could be used to examine the degree to which specific activities impose demands on areas of the brain that regulate mental arithmetic and processes like working memory and executive control in unimpaired individuals and clinical groups.

9.8 Summary

Different scientific fields, such as psychology, mathematics, and history, have all been incorporated into the science of mathematics education. Cognitive neuroscience provides a set of techniques, methodologies, and theories for studying cognitive processes that occur during mathematical thinking, which may supplement and extend knowledge gained from behavioral data. This chapter discussed cognitive neuroscience in education, specifically in mathematics. It discussed developmental cognitive neuroscience's implications for learning and teaching for academic accomplishment. This chapter also examines the numeracy and reasoning of mathematical learning in cognitive neuroscience.

CHAPTER 10

ISSUES AND CHALLENGES

10.1 Introduction

Several values are shared by mathematics and science education, and they face the same difficulties and challenges. However, there are distinctions between these two fields of study, necessitating the creation of two independent manuals. For instance, unlike science, it is undeniable that mathematics must be taught to all students from the start of obligatory school. Mathematics education is not always offered to students' satisfaction, but it is available to all students attending school. Although it is widely acknowledged that mathematics should be taught in primary school, this does not mean that mathematics education is without controversy. According to nationally and internationally assessments, many kids' mathematics understanding, and competencies fall short of the desired level after primary education.

Furthermore, the differences seen between and within countries are alarming. Even among students who receive excellent assessment results, some dislike mathematics and do not see the value in devoting so much of their school time (*Challenges in Basic Mathematics Education.* 2012). These conclusions suggest that perhaps the objectives outlined in the preface are still a long way off and that the high number of young people without access to school was not the only barrier to their success, even though it is a significant one.

Students should create a positive and acceptable picture of mathematics through quality teaching. For this to be feasible, it must be loyal to mathematics, both in terms of substance and procedures. It should help students comprehend which requirements are satisfied by the mathematics they are taught, as well as the fact that mathematics has a vast history intertwined with humankind's history. Learning mathematics requires obtaining the skills necessary to explore this historical tradition. Mathematics learning should thus teach students to recognize that mathematics is not a static accumulation of research but rather a dynamic and developing discipline whose growth is sustained by other scientific fields and, in turn, nurtures them. This should help students perceive mathematics as a profession that must assist in solving today's critical global issues, as indicated in the joint introduction (Register, Stephan, and Pugalee, 2021). Thus, a concept of mathematicians as a dynamic subjects engaging with the actual world, open to interactions with other professions, and not constrained by research domains must underpin quality mathematics teaching.

One of the most critical problems in teaching mathematics is individuals' mathematical proficiency. Despite this, many scholars consider mathematics one of the most difficult fundamental courses to master. This pessimistic thought might be caused by various reasons that obstruct their math education. This study investigated the challenges, hurdles, and challenges encountered by learners in the mathematics learning process to understand better the factors that impede their learning. Because mathematics is linked to many other professions and sciences, it has traditionally received considerable emphasis in education.

Furthermore, student performance in mathematics has been a hot topic and is considered a worldwide issue in many countries. Arithmetic issues have been linked to a lack of regulatory skills within pupils in learning math and being considered a complex topic. Self-regulation is a comprehensive concept that encompasses the stages of learning beforehand, during, and after. Self-regulation in studying is linked to 21st-century learning competency; therefore, students who are not regulated will have a tough time overcoming obstacles or hurdles during learning (Shumway 1980).

Students are significantly affected by the frequent changes in the educational system and delivery techniques. This condition necessitates kids learning more independently and effectively. To accomplish this, pupils must be taught how to improve their ability to find the most appropriate learning technique. Failure to do so will impact students' learning motivation and may eventually cause them to lose interest in learning. Motivation is a crucial part of the learning process since it serves as an activator and motor for completing a task satisfactorily. As a result, motivation is essential for an individual to effectively-identified and document obstacles.

Furthermore, students will use incentives to explain or predict their training and performance behavior. Academic motivational behaviors such as the desire to complete tough work and stay longer in challenging settings will determine students' confidence to confront everyday school life problems (McLeod 1992). Consequently, the incentive is required for people to identify and document difficulties successfully. In addition, learners will use motivation to explain or forecast their coaching and development performance. Academic motivating behaviors such as the desire to finish difficult work and stay longer under challenging situations will impact students' comfort in dealing with issues in their ordinary school career.

10.2 Basic Challenges and issues in Mathematics Education

Education, overall, and mathematics participation in education have constantly confronted and will continue to experience obstacles: it could hardly progress without overcoming adversity, some of which date back thousands of years, yet when complaints were made about the challenges of learning and school staff made complaints about students' lethargy. To be accurate, each epoch brings its unique set of issues, approaches to solving old and new difficulties, and achievements. At the same time, certain issues may be considered global, not specific to any one location (even if they present themselves differently in different parts of the world), while others are indigenous to specific countries and regions. In mathematics education, there are several theories and ideologies. The two ideologies and concepts that were widely argued and discussed in detail in mathematics education are radicalism and sociological structuralism (Freudenthal 1981). Most mathematics instructors and curricular specialists' worldviews have been influenced by conceptions of mathematics, such as mathematics as a foreign subject, mathematicians as a set of signs, maths as a useless topic, math as a body of actual knowledge, and mathematics as objective knowledge.

The majority of definitional concerns are laborious and tedious. Nonetheless, for an academic or professional debate to be serious, if not possible, it is necessary for individuals involved to understand at least the fundamental objects and conceptions. This is especially true in an area like teaching mathematics, where simplicity and precision are challenging to attain, let alone accept as a given. Researchers understand a cause for offering mathematics education to students within some section of the educational system as a driving force, typically of a general nature, that has inspired and given rise to the presence of mathematics teaching within that segment, as determined by the bodies that make decisions (including non-decisions) in the system at issue.

10.3 Difficulties in Mathematics

The international comparative results of both TIMSS and PISA investigations demonstrate that school children have roughly similar mathematical knowledge. The scores have one of the minor standard deviations among the countries examined. Furthermore, the inequalities between males and females, as well as between schools, are minor. However, there are significant gaps in education and basic abilities between Finns' older and younger generations. Moreover, a sizable proportion of young individuals do not pursue higher education. These difficulties were considered based on the survey of Finnish employment (Banerjee and Seshaiyer 2019). These issues are exacerbated in older persons; their knowledge may be significantly lower than that of younger Finnish adults. Adults with poor basic arithmetic skills or who have previously struggled with mathematics achievement are frequently unable to pursue their education or drop out of post-secondary school.

10.3.1 Impact of Textbook

The effects of textbooks on the content of mathematics classes and teachers' teaching practices attracted enormous attention to all the topics on which effects have been investigated. Using a mixed-methods design that included teaching methods, interviews, and questionnaires, the researcher evaluated the influence of volumes on the substance of instruction in three separate instructors' Slovenian mathematics sessions. The textbook influence was assessed using three categories: "textbook absent," "textbook direct," and "textbook indirect."

The impact of textbooks on students' mathematical achievement is a critical question in the field of textbook impact. Törnroos addresses learning opportunities in three ways: 1) The percentage of each TIMSS topic covered in each volume offered an initial estimate of learning chances. 2) The second method was based on a teacher question, in which teachers were asked when and if they had delivered the TIMSS themes. 3) An item-based study of textbooks produced the third measure of learning opportunities. In the latter approach, various textbooks were evaluated based on the preparation level they gave students to respond to the TIMSS 1999 items.

10.3.2 Culture Challenge and issues in Mathematics

One might approach this from various angles, such as concentrating on the education of ethnically diverse pupils or analyzing the condition of migrant instructors who have received their education in another nation. It appears that, following a widespread belief held by many individuals, mathematics education, like mathematics itself, is seen as ubiquitous and civilization and value-free. This is not to say that they believe mathematics has no significance; instead, they believe that mathematics learning does not impart or derive from any values other than those that a group or society promotes through its academic institutions. Another aspect of mathematical activity that almost everyone is aware of is management's importance. It entails appreciating the presence and stability that arithmetic provides through having rules, knowing the ability to foresee, and being eligible to utilize concepts in real-world circumstances. Progress is a value that is complementary to control (Jebson 2012). Mathematicians believe they can explore and build things based on this understanding since it feels comfortable. Mathematical progress is about change and alternatives; it is attained by looking for instances in which generalization fails. The hunt for the counter-example, the extreme instance, and logical discrepancies create new issues to solve.

Table 10.1 Classification of values based on observation

		Observed		
		Taught implicitly	Taught explicit	Not Implemented
Nominated/ Premeditated	Nominated Explicit	Co-operation	Self-esteem	Creativity
	Not nominated	Individual dissimilarity	Exhaustiveness	-

According to our classroom studies, various values were being formed, but the project team was more interested in seeing how much control the teachers had over their value teaching. We could categorize whether teachers did or did not nominate the qualities that were later seen based on examining the case-study work of the research project (or sometimes not observed) (Panthi and Belbase 2017). We classified the proposed values as explicit or implicit when teachers were observed teaching them, which is summarized in Table 10.1: Classification of values based on observation.

Based on the cognitive approach, with its concentration on individuals' cognitive and intra-individual characterizations and explanations, which has tended to disregard the socio-cultural solid background of mathematics learning, the social-economic approach to research on mathematics learning is critical. However, social-economic studies in teaching mathematics have focused mainly on learning within specific cultural practices and groups, overlooking two critical factors. To begin with, the focus on intercultural practice differences has hidden significant cross distinctions within those cultural groupings. Second, research has tended to overlook the parts of learning that occur during the transition between various cultural practices (Christopoulos et al., 2020).

In a social outcast population combination, teachers and students frequently cause communication disruptions by merely engaging in routine and practical everyday activities in ways that respective sub-cultures define as acceptable and appropriate... The issue is a cultural clash, not "stupid students" or "racist teachers."

10.4 Teaching Practices

The challenges, as well as adjustments in instructional procedures, must be made following the stated objectives. This is not commonly the situation, according to investigations on classroom practices of education and training studies and surveys performed by multilateral institutions (European Commission, 2007). Basic mathematics instruction is still far too often tedious for the following reasons:

- It is meant as formal education, with a focus on learning procedures and memorizing rules with no apparent justification to the students;
- Pupils who are unaware of those requirements are satisfied by the mathematical topics presented or how they are linked to previously learned ideas.
- The linkages to the actual world are poor, the implementations are stereotyped, and the links to the real world are often too articled to be credible.
- There are only a few experimental and simulation exercises.
- Technologies are rarely applied in a meaningful way;
- In their mathematics work, students have little flexibility and frequently repeat exercises.

In a rising series of studies and experiments over the decades, it has been demonstrated that alternate learning strategies exist, giving students a new perspective on mathematics and their ability to grasp its significance. Such alternative solutions are typically based on formwork learning methods and emphasize the issue's importance in teaching and learning mathematics. Those troubles are being used for student engagement and preparing them for the implementation of novel constructs or to facilitate them to study and apply the theories that have already been tried to introduce. Since mathematical education is viewed as a gradual transition, in which significance is built up by trying to compare, going to meet carefully chosen complex issues, and trying to draw on wide and varied processes of interpretation and equipment, learning is viewed as a gradual transition in which meaning is built up by trying to compare meeting carefully chosen complex issues and drawing on diversified processes of formal recognition and gadgets (Schleppegrell 2007).

This highlights the issue of teacher training and the tools available to help them improve their methods. This topic will be discussed further in this text. However, some marks should indeed be made now: in specific, practical professional learning should assist educators in the software design documents that will allow for mathematically constructive interrogations within the restrictions of the classroom, as well as in playing their role as guide and mediator in managing these tasks in a mathematically effective manner (Panthi and Belbase 2017). Changes in practices must also be examined dynamically, with received attention to maintain a sufficient distance between new and old practices, and therefore must be consistently supported by sufficient resources to start and maintain the required reforms. Coaches are frequently provided with models of practice that are too distinct from their own experiences to be integrated without becoming perverted, whether in pre-service education or once materials are made available to them.

In addition, new methods' increased mathematical and instructional skill is grossly overestimated. This makes it difficult for teachers to see the benefits of suggested modifications, and they are thus unmotivated to implement them.

According to our classroom studies, various values were being formed, but the project team was more interested in seeing how much control the teachers had over their value teaching. We could categorize whether teachers did or did not nominate the qualities that were later seen based on examining the case-study work of the research project (or sometimes not observed) (Panthi and Belbase 2017). We classified the proposed values as explicit or implicit when teachers were observed teaching them, which is summarized in Table10.1 Classification of values based on Observation.

Based on the cognitive approach, with its concentration on individuals' cognitive and intra-individual characterizations and explanations, which has tended to disregard the socio-cultural solid background of mathematics learning, the social-economic approach to research on mathematics learning is critical. However, social-economic studies in teaching mathematics have focused mainly on learning within specific cultural practices and groups, overlooking two critical factors. To begin with, the focus on intercultural practice differences has hidden significant cross distinctions within those cultural groupings. Second, research has tended to overlook the parts of learning that occur during the transition between various cultural practices (Christopoulos et al., 2020).

In a social outcast population combination, teachers and students frequently cause communication disruptions by merely engaging in routine and practical everyday activities in ways that respective sub-cultures define as acceptable and appropriate... The issue is a cultural clash, not "stupid students" or "racist teachers."

10.5 Issues in Mathematics

One of the challenges in reforming the course content in the country is a lack of adequate theory-building of issues that influence curriculum creation, learning, and teaching of a subject. For the sake of directing these reform measures, domestic research in the field of mathematics education is insufficient. Much of such research, mainly in the psychological paradigm, has tackled content-related problems in the primary grades. They do not constantly interact with any concept of teaching mathematics that could help us better understand why anything succeeds or fails, and they do not provide us with adequate knowledge of students' thinking and problem-solving processes (Bashmakov 2010). Only in the last few years have specific individuals and a few institutions made severe attempts to comprehend challenges with learning and teaching various aspects of math, revealing children's processes of coming at answers to problems in the classroom instruction environment. This is also primarily concerned with elementary school students and is insufficient to drive improvements in secondary and senior secondary schools. These have not yet been influential enough to flow into curricular reform due to several constraints of such research.

Recognizing how children learn common content areas is not enough to justify a new curriculum. Additionally, essential ideas must be addressed. A discussion and analysis of the reasons for mathematics teaching in primary, secondary, and senior secondary schools and how they are related to the country's variety and the draws and pushes of various significant sections of the population. We need to grasp better the context and character of a child's teaching mathematics. Researchers must also participate with concepts such as characterizations and signifiers in mathematical concepts, things to make, language problems, rationale, logical arguments, and trying to prove, use of technology, understanding educational cultures, teacher preparation, cultural issues, and their impact on teaching mathematics, actually impact and mathematical skills education-learning, and evaluation. These concerns directly impact the framework we use to construct a curriculum (Smith 1996). To accomplish this, designers will need the assistance of many more individuals and institutions. Nonetheless, many people believed that such mathematics was also not very practical and that the applications in terms of word problems were a bit clumsy. Another approach to making mathematics meaningful was incorporating mathematics into real-world contexts for children, interacting with critical learning philosophies and methodologies.

10.6 Issues in Elementary Education

Primary arithmetic textbooks included various real-life stories within chapters and established specific thematic chapters to address various topics such as work, business, culture, trade expertise, and the background of cave paintings, among others. According to research, students can display significantly deeper mathematics comprehension, and their achievement on conventional mathematical metrics is also more vital when issues are based on learners' contexts. Even if most chapters focused on topics that were gradually developed as outlined in the syllabus, a few special themed sections used circumstances that invoked and combined topics previously learned. For example, building with Bricks, The Junk Seller, and The Fish Tale was theme units that combined forms, numbers, measurements, currency, pattern, and other ideas through natural environments. Some of them were created to situate arithmetic in a social-economic context, based on challenges raised by masonry, masonry architecture, brick manufacturing craftspeople, trash collectors and sellers, and fish-workers, boatsmen, and fish dealers.

Furthermore, the reactionary curriculum allows for a purely symbolic strategy in which "joyous" multicultural portrayals are limited to perceiving diversity through the prism of an integral part of culture "other," with no critical dialogue concerning concerns of "different," prejudice, or supremacy (Silver and Kilpatrick 1994). It necessitates differentiating among real-world mathematics aspects, such as purchasing, visiting, teaching, or constructing, and those that urge pupils to critically analyze issues of inequality through a sense of shared social agency.

Evidently, throughout a century of syllabus recertification attempts in Kerala, only one state in the country to behave accomplished nearly unanimous primary education and educational achievement while still particularly bothered regarding issues of quality and equal education, a significant difference in the sequence of assessments was affected, closely tied with schoolbook and pedagogies reforms. Significantly, every query, or 'carry out an assessment activity,' as it was dubbed, was regarded as a learning activity, keeping the action perspective to classroom interactions (Bishop 2002). Instructors have low standards for these marginalized kids and profound social biases toward them. The conventional assumptions they hold about the learner–as someone with whom education is vital, with relatively close punctuality, but whose parents arrange for individualized attention–may not matches the learners they experience in the classroom. Special educators have little access to help them cope with it and develop the peculiarities that pupils learn in the classroom. Without substantial steps to educate educators, give them support to deal with the variety of difficulties that denigration brings to schools, and assist them in appreciating the NCF approach to addressing equity and social justice concerns in mathematics.

This necessitates a robust teacher education program seeking to address every one of these challenges. Regrettably, teacher preparation organizations have been underestimated, ignored, and rendered invisible in the more extraordinary academic landscape, with very little support from higher education institutions, especially research organizations. There is an immediate need for such a sovereign to position a significant focus on teacher development, to deliberately detect within the research and academic agendas of universities and organizations, to reinvigorate teacher learning institutions, and rewrite their education system so that they could provide necessary in-service and which was before instruction for educators to conduct business the new curriculum framework in mathematic.

10.7 Issues in Secondary Education

An issue, by definition, is a task whereby the individual encountering it:

1. Needs or wants to come up with a solution
2. Has no readily available technique for finding the solution
3. Must make some attempt to discover a solution

Modeling is similar to solving problems in so many aspects. Students are provided real-life events, and they are asked many questions about them. The situations should pique the pupils' curiosity, prompting them to seek answers. On the other hand, students cannot tackle the problem directly since the mathematics underlying a real-world modeling work is often hidden (Pia 2015). Finally, because modeling work is a factor in students' in-class obligations, an attempt to answer the issue must be undertaken.

To summarize, modeling gives a problem set in which theoretical concepts can be developed. As a result, it seems reasonable to believe that modeling in the classroom would aid in the impact on pupils' issue abilities. Indeed, a study using modeling as a training tool backs up this assertion.

"Integration" is a hot topic among teachers nowadays. As high school teachers, researchers should emphasize linkages between mathematics and science, art, political science, and other aspects of the program. Additionally, students should express their mathematical accomplishments, knowledge, and understanding in writing and conversational form. The ideas offered by English language teachers are repeated as a result. Simulation, almost by essence, emphasizes the linkages between mathematics and other subjects. Furthermore, the intricacy of the modeling process, and several teachers' collaborative attitudes to modeling, encourage class discussion. Learners might be challenged to explain the relationship between their models and the starts with problem scenarios, the procedures by which their models were produced, the applications of their models, and their constraints as they develop their models and engage with each other (Asikhia 2014).

10.8 Issues in Senior Secondary Education

Conflict resolution is approached in two ways in mathematics teaching. (i) As a study topic on questions such as: How is conflict resolution linked to other parts of "mathematical knowledge"? What are the most crucial intrinsic and cognitive elements of solving problems? How would one properly categorize various issue processes? What are the most significant affective and cognitive barriers to pupils' achievement of issue skills? Is crisis management something that can be taught and learned? (ii) In math education, questions surrounding the inclusion and implementation of problem-solving in mathematics curricula are addressed. The essential issues and difficulties regarding modeling, applied problem solving, implementations, and the interaction among mathematics and topics (Panthi and Belbase 2017). Given the scope and constraints of this article, it will not be feasible to provide an entire list of the significant features, nor would it be feasible to provide a thorough and detailed explanation of those presented. We will have to limit ourselves to summarizing a few key topics here. The emphasis will be on curriculum aspects, with less focus on theory in general.

10.9 Challenges in Mathematics Education

"The most important resource for teaching mathematics is textbooks. According to TIMSS 2007, textbooks are used as the major basis for mathematics lessons by 65 percent of Upper primary teachers and 60 percent of Grade 8 teachers around the world." This is one of the primary findings of a Nuffield Foundation-sponsored comparison study of global teaching mathematics. It demonstrates the need for textbooks in the teaching and learning process in general education, particularly in this age of digital information systems. Even if confined to the spectrum of textbooks used among teachers, textbooks must consider one key caveat: "use of textbooks" has numerous meanings. Assessing the development of a textbook for first-year university mathematics students appears to be a unique instance. Using an online application, users were ready to communicate the textbook and make changes to the material. In terms of textbook use in schools, he identifies four types of "use" by using the terms "textbook" and "curriculum" interchangeably.

The sheer identification of these various conceptions of use indicates that evaluating textbooks used will be complex. The concept of the school curriculum and its "usage" underpins methodological techniques. In our community mathematics, a sociopolitical approach differentiates between the artifact textbook, the user (whether students or teachers), and the object. The latter, in particular, provide a solution to the problems revealed by the various approaches, as they integrate the benefits of various methods while minimizing their shortcomings, seeking to address validity issues through triangulation (Smith 1996). The researcher incorporates information from a survey question, a question-and-answer session with the teacher, direct observation of lecture notes and task-solving meetings, mock interviews, and some informal communication with students to recognize educators' strategies for mathematics teaching and the constraints and opportunities that impact them. The method selection is well thought out and guided by the goal of reducing difficulties with each approach to improving the neutrality and authenticity of the information.

In most cases, studies in teaching mathematics are more concerned with the impact of the use on the user, therefore addressing the survey of what effect books have on students. In this study, researchers use the term "impact" rather than "effect" to separate it from the scientific phrase "effect size," which is now a frequent statistic in social sciences (particularly meta-analyses).

Professional numeracy is currently defined as the 'basics' of the four computing operations, as well as components of algebraic, geometric, statistical thinking or quantitative literacy, and problem-solving, according to the present investigation. Furthermore, "unofficial and non-standard mathematical practices and processes in adults' life, which may have little similarity to conventional, taught math concepts" are recognized as being involved. Mature mathematics is more complicated than 'applied' or 'practical' mathematics since it draws on skills and information acquired over a lifetime (Hohenwarter, Hohenwarter, and Lavicza 2009). No extensive studies explain the situation of adults' basic math skills, yet adults' challenges in math have been seen regularly in vocational education. Mathematics teachers have found that modern kids' mathematical abilities are lower than often supposed. One explanation is the dearth of numeracy material in today's elementary school mathematics textbooks. Students in school do not receive a solid foundation in computation.

10.10 Summary

The chapter describes the challenges and issues involved in mathematics education. In the first section, the essential introduction to mathematic education based on its issues and challenges was expressed and continued to the fundamental challenges and issues involved. Each section describes the issues in the education system based on the textbook, culture, and other impacts on the mathematics education system. Finally, the chapter concluded with the challenges and benefits of the mathematics education system.

REFERENCES

1. Ainley, J. Pratt, D., & Hansen, A. 2006. Connecting engagement and focus in pedagogic task design. British Educational Research Journal, 32(1), 23-38. Retrived and downloaded date 12/03/ 2014
2. Akangbou, S.D. 1984. Planning the junior secondary school curriculum. In Pai Obanya (edited). Curriculum In Theory and in practice. Ibadan: Educational Research and study group
3. Algani, Y. 2018. Applying creative skills in teaching math at the primary school stage. Journal of International Economy and Business, 6, 26-33.
4. Ardoin, Scott P., et al. 2005. "Application of a three-tiered response to intervention model for instructional planning, decision making, and the identification of children in need of services." *Journal of Psychoeducational Assessment* 23.4: 362-380.
5. Ariani, Y, Y Helsa, S Ahmad, and Rci Prahmana. 2017. "Edmodo Social Learning Network for Elementary School Mathematics Learning." *Journal of Physics: Conference Series* 943 (December): 012056. https://doi.org/10.1088/1742-6596/943/1/012056.
6. Asikhia, OA. 2014. "Effect of Cognitive Restructuring on the Reduction of Mathematics Anxiety among Senior Secondary School Students in Ogun State, Nigeria." *International Journal of Education and Research* 2 (2): 1–20.
7. Attard, Catherine. 2010. "Students' Experiences of Mathematics during the Transition from Primary to Secondary School." *Mathematics Education Research Group of Australasia.*
8. Bamus, G.A. 2002. Curriculum development and changes in mathematics education in Nigeria. A paper presented during the mathematics education Summit at National Mathematical Centre, Abuja.
9. Banerjee, Rakhi, and Padmanabhan Seshaiyer. 2019. "Challenges in School Mathematics Curriculum Reform in India: Transforming Teacher Practices Through Pedagogical Innovations." In *School Mathematics Curricula*, edited by Catherine P. Vistro-Yu and Tin Lam Toh, 227–37. Mathematics Education – An Asian Perspective. Singapore: Springer Singapore. https://doi.org/10.1007/978-981-13-6312-2_12.

10. Bashmakov, Mark. 2010. "Challenges and Issues in Post-Soviet Mathematics Education." *Russian Mathematics Education: History and World Significance* 4: 141.
11. Bennett, David A., et al. 2006. "The effect of social networks on the relation between Alzheimer's disease pathology and level of cognitive function in old people: a longitudinal cohort study." *The Lancet Neurology* 5.5: 406-412.
12. Belsito, Courtney. 2016. "The importance of 'teacher quality and 'quality teaching on academic performance." *Journal of Student Engagement: Education Matters* 6.1: 28-38.
13. Bishop, Alan J. 2002. "Critical Challenges in Researching Cultural Issues in Mathematics Education." *Journal of Intercultural Studies* 23 (2): 119–31. https://doi.org/10.1080/07256860220151041.
14. Bishop, Alan J. 1988. "Mathematics education in its cultural context." *Educational studies in mathematics* 19.2: 179-191.
15. Bodovski, Katerina, and George Farkas. 2007. "Mathematics Growth in Early Elementary School: The Roles of Beginning Knowledge, Student Engagement, and Instruction." *The Elementary School Journal* 108 (2): 115–30. https://doi.org/10.1086/525550.
16. Bukatku, D & Daehler, W. 1995. Child Development: A Thematic Approach. (2nd edt) New Jersey.houghton miffin Company.
17. Callahan, Leroy G., and Vincent Joseph Glennon. 1975. *Elementary School Mathematics: A Guide to Current Research*. 4th ed. Washington: Association for Supervision and Curriculum Development.
18. Carey, S. 2009. The origin of concepts. New York, NY: Oxford University Press.
19. *Challenges in Basic Mathematics Education.* 2012. Paris: UNESCO.
20. Charles-Ogan, Gladys, and Nchelem Rosemary George. 2015. "INVESTIGATING DIFFICULT CONCEPTS IN SENIOR SECONDARY SCHOOL MATHEMATICS CURRICULUM AS PERCEIVED BY STUDENTS" 3 (6): 8.
21. Chatzistamatiou, Mariza, Irini Dermitzaki, Anastasia Efklides, and Angeliki Leondari. 2015. "Motivational and Affective Determinants of Self-Regulatory Strategy Use in Elementary School Mathematics." *Educational Psychology* 35 (7): 835–50. https://doi.org/10.1080/01443410.2013.822960.
22. Christopoulos, Athanasios, Henri Kajasilta, Tapio Salakoski, and Mikko-Jussi Laakso. 2020. "Limits and Virtues of Educational Technology in Elementary School Mathematics." *Journal of Educational Technology Systems* 49 (1): 59–81. https://doi.org/10.1177/0047239520908838.

23. Ciftci, S Koza. 2015. "Effects of Secondary School Students' Perceptions of Mathematics Education Quality on Mathematics Anxiety and Achievement." *Educational Sciences: Theory and Practice* 15 (6): 1487–1501.
24. Civil, Marta. 2007. "Building on community knowledge: An avenue to equity in mathematics education." *Improving access to mathematics: Diversity and equity in the classroom*: 105-117.
25. Clarke, D., Goos, M., & Morony, W. 2007. Problem solving and working mathematically: An Australian perspective. ZDM Mathematics Education, 39, 475–490.
26. Cobb, P. & Hodge, L. L. 2002. A relational perspective on issues of cultural diversity and equity as they play out in the mathematics classroom. Mathematical Thinking and Learning, 4, 249-284.
27. d'Ambrosio, Ubiratan. "Socio-cultural bases for mathematical education." *Proceedings of the fifth international congress on mathematical education*. Birkhäuser, Boston, MA, 1986.
28. Dean, Peter G. 2019. *Teaching and Learning Mathematics*. Routledge.
29. Department of Children, Schools and Families [DCSF] 2008. Secondary mathematics guidance papers. Nottingham, UK: DCSF.
30. De Smedt, Bert, Daniel Ansari, Roland H. Grabner, Minna M. Hannula, Michael Schneider, and Lieven Verschaffel. 2010. "Cognitive Neuroscience Meets Mathematics Education." *Educational Research Review* 5 (1): 97–105. https://doi.org/10.1016/j.edurev.2009.11.001.
31. Dubinsky, Janet M., Gillian Roehrig, and Sashank Varma. 2013. "Infusing neuroscience into teacher professional development." *Educational Researcher* 42.6: 317-329.
32. Emron, S. & Dhindsa, H.S. 2010. Integration of Interactive White Board Technology to improve secondary science teaching and learning.
33. English, Lyn D., and Graeme S. Halford. *Mathematics education: Models and processes*. Routledge, 2012.
34. Ernest, Paul. *Mathematics Teaching: The State of the Art*. The Falmer Press, Taylor & Francis, Inc., 1900 Frost Rd., Suite 101, Bristol, PA 19007, 1989.
35. Ernest, Paul. 1985. "The philosophy of mathematics and mathematics education." *International Journal of Mathematical Education in Science and Technology* 16.5: 603-612.
36. Fatade, Alfred O, Abayomi A Arigbabu, David Mogari, and Adeneye OA Awofala. 2014. "INVESTIGATING SENIOR SECONDARY

SCHOOL STUDENTS' BELIEFS ABOUT FURTHER MATHEMATICS IN A PROBLEM-BASED LEARNING CONTEXT." *Bulgarian Journal of Science and Education Policy* 8 (1): 5.

37. Federal Ministry of Education, 2007. 9-year basic education curriculum for Primary 1 – 3. Abuja: NERDC

38. Fisher, K. D. 2010. Technology-enabled active learning environments: An appraisal. *CELE Exchange. Centre for Effective Learning Environments, 2010*(6–10), 1–8.

39. Fischer, Kurt W. 1987. "Relations between brain and cognitive development." *Child development:* 623-632.

40. Freudenthal, Hans. 1981. "Major Problems of Mathematics Education." *Educational Studies in Mathematics* 12 (2): 133–50.

41. Geake, John, and Paul Cooper. 2003. "Cognitive Neuroscience: Implications for Education?" *Westminster Studies in Education* 26 (1): 7–20. https://doi.org/10.1080/0140672030260102.

42. Ginsburg, Herbert P. 2009. "The challenge of formative assessment in mathematics education: Children's minds, teachers' minds." *Human development* 52.2: 109-128.

43. Gopnik, A., Wellman, H.M., 2012. Reconstructing constructivism: causal models, Bayesian learning mechanisms, and the theory theory. Psychological Bulletin 138 (6), 1085–1108.

44. Gravemeijer, Koeno Pay Eskelhoff. 1994. "Developing realistic mathematics education".

45. Greer, Brian, et al., eds. *Culturally responsive mathematics education*. Routledge, 2009.

46. Gyöngyösi Wiersum, Erika. 2012. "Teaching and Learning Mathematics through Games and Activities." *Acta Electrotechnica et Informatica* 12 (3). https://doi.org/10.2478/v10198-012-0026-2.

47. Halmos, Paul R. "Applied mathematics is bad mathematics." *Mathematics tomorrow*. Springer, New York, NY, 1981. 9-20.

48. Hamilton, D. 1977. Making sense of curriculum evaluation: continuities and discontinuities in an educational idea, in Deakin University, 1987, Course Reader, 1: Approaches and Dilemmas in Curriculum Evaluation, Melbourne, Australia.

49. Harel, Guershon, and Jana Trgalova. "Higher mathematics education." *International handbook of mathematics education*. Springer, Dordrecht, 1996. 675-700.

50. Harris, P.L., 2002. What do children learn from testimony? In: Carruthers, P., Siegal, M., Stich, S. (Eds.), The Cognitive Basis of Science. Cambridge University Press, New York, pp. 316–334.

51. Hinton, Christina, and Kurt W. Fischer. 2010. "Learning from the developmental and biological perspective." *The nature of learning: Using research to inspire practice*: 113-34.
52. Hossain, Anowar, and Rohani Ahmad Tarmizi. 2013. "Effects of Cooperative Learning on Students' Achievement and Attitudes in Secondary Mathematics." *Procedia-Social and Behavioral Sciences* 93 (October): 473–77. https://doi.org/10.1016/j.sbspro.2013.09.222.
53. Hohenwarter, Judith, Markus Hohenwarter, and Zsolt Lavicza. 2009. "Introducing Dynamic Mathematics Software to Secondary School Teachers: The Case of GeoGebra." *Journal of Computers in Mathematics and Science Teaching* 28 (2): 135–46.
54. Ifamuyiwa, S.A., and M.K. Akinsola. 2008. "Improving Senior Secondary School Students' Attitude towards Mathematics through Self and Cooperative-Instructional Strategies." *International Journal of Mathematical Education in Science and Technology* 39 (5): 569–85. https://doi.org/10.1080/00207390801986874.
55. Israel Ministry of Education. 2018. Curriculum of science and technology, Jerusalem-Israel (Hebrew).
56. Jacobson, Michael J., and Robert B. Kozma, eds. *Innovations in science and mathematics education: Advanced designs for technologies of learning*. Routledge, 2012.
57. James, Ajogbeje Oke, and Alonge Micheal Folorunso. 2012. "Effect of Feedback and Remediation on Students' Achievement in Junior Secondary School Mathematics." *International Education Studies* 5 (5): p153. https://doi.org/10.5539/ies.v5n5p153.
58. Jaudinez, Aljemedin Sakilan. 2019. "Teaching Senior High School Mathematics: Problems and Interventions." *Pedagogical Research* 4 (2). https://doi.org/10.29333/pr/5779.
59. Jebson, Sofeme R. 2012. "Impact of Cooperative Learning Approach on Senior Secondary School Students Performance in Mathematics." *Ife PsychologIA*, 7.
60. Johnson, Mark H., and Yuko Munakata. 2005. "Processes of change in brain and cognitive development." *Trends in cognitive sciences* 9.3: 152-158.
61. Jonassen, D. H., Peck, K. L., & Wilson, B. G. 1999. Learning with Technology: A Constructivist Perspective, New Jersey: Merrill/Prentice Hall.
62. Jong, Ton de, Tamara van Gog, Kathleen Jenks, Sarah Manlove, Janet van Hell, Jelle Jolles, Jeroen van Merriënboer, Theo van Leeuwen, and Annemarie Boschloo. 2009. *Explorations in Learning and the Brain*. Boston, MA: Springer US.

https://doi.org/10.1007/978-0-387-89512-3.

63. Kaur, B. 2001. TIMSS & TIMSS-R – Performance of grade eight Singaporean students. In C. Vale, J. Horwood, & J. Roumeliotis (Eds), 2001 a mathematical odyssey (pp. 132-144). Proceedings of the 38th annual conference of the Mathematical Association of Victoria, Brunswick, Vic: MAV.
64. Keitel, Christine, et al. 1989. "Mathematics, education, and society." *Science and Technology Education, Document Series* 35.
65. Kendra Cherry. 2014, Piaget's Stages of Cognitive Development. [Accessed: 23 September 2014 available at. http://psychology.about.com/od/piagets theory/a/keyconcepts.html.
66. Kilpatrick, J., Swafford, J., & Findell, B. (Eds.). 2001. Adding it up: Helping children learn mathematics. Washington, DC: National Academy Press.
67. Krainer, Konrad. "Innovations in mathematics, science and technology teaching." *Connected Maths. MAV Annual Conference*. 2008.
68. Kundu, Arup, and Aditi Ghose. 2016. "The Relationship between Attitude towards and Achievement in Mathematics among Higher Secondary Students." *International Journal of Multidisciplinary Research and Development* 3 (6): 69–74.
69. Kyriacou, Chris. 1992. "Active Learning in Secondary School Mathematics." *British Educational Research Journal* 18 (3): 309–18. https://doi.org/10.1080/0141192920180308.
70. Lazarus, S. 2010. Educational Psychology: in social context.4th edition. Cape Town. Oxford University Press
71. Lee, Hon Wah, and Chi-Hung Juan. 2013. "What Can Cognitive Neuroscience Do to Enhance Our Understanding of Education and Learning?" *Journal of Neuroscience and Neuroengineering* 2 (4): 393–99. https://doi.org/10.1166/jnsne.2013.1064.
72. Lesh, Richard, and Anthony E. Kelly. 1997. "Teachers' evolving conceptions of one-to-one tutoring: A three-tiered teaching experiment." *Journal for Research in Mathematics Education* 28.4: 398-430.
73. Longe, R.S. 1984. Planning for a Responsive curriculum. In Pai Obanya edited. Curriculum, in theory and in practice Ibadan: Educational Research and study group.
74. Mahmood, Sadia. 2011. "Development and Validation of the Mathematics Anxiety Scale for Secondary and Senior Secondary School Students," 12.

75. Malik, A. B. 2010. Public–private partnerships in education: Lessons learned from the Punjab education foundation. Asian Development Bank.
76. Marsh, C.J., & Willis, G. 2007. Curriculum: alternative approaches, ongoing issues (4thed.). New Jersey, NJ: Pearson/Merrill Prentice Hall.
77. McLeod, Douglas B. 1992. "Research on Affect in Mathematics Education: A Reconceptualization." *Handbook of Research on Mathematics Teaching and Learning* 1: 575–96.
78. Menon, Vinod. 2010. "Developmental Cognitive Neuroscience of Arithmetic: Implications for Learning and Education." *ZDM* 42 (6): 515–25. https://doi.org/10.1007/s11858-010-0242-0.
79. Mishra, L. 2007. Teaching of mathematics. New Delhi: APH Publication
80. Miyazaki, Mikio. 2013. "Levels of Proof in Lower Secondary School Mathematics," 22.
81. Montague, Marjorie, and Brooks Applegate. 1993. "Mathematical Problem-Solving Characteristics of Middle School Students With Learning Disabilities." *The Journal of Special Education* 27 (2): 175–201. https://doi.org/10.1177/002246699302700203.
82. Moyosore, Olagunju Asimiyu. 2015. "THE EFFECT OF FORMATIVE ASSESSMENT ON STUDENTS' ACHIEVEMENT IN SECONDARY SCHOOL MATHEMATICS" 3 (10): 10.
83. Moore, Robert C. 1994. "Making the transition to formal proof." *Educational Studies in mathematics* 27.3: 249-266.
84. National Board of Higher Mathematics (n.d.). Retrieved April 9, 2012 from: http://www.nbhm.dae.gov.in/about.html.
85. National Council of Educational Research and Training. 2006. National focus group on aims of education. New Delhi: NCERT.
86. National Council for Educational Research and Training. 2005. National curriculum framework. Delhi: NCERT publications.
87. Ng, Lay Keow. 2012. "Mathematics Anxiety in Secondary School Students." *Mathematics Education Research Group of Australasia*.
88. Niss, Mogens. "Assessment in mathematics education and its effects: An introduction." *Investigations into assessment in mathematics education*. Springer, Dordrecht, 1993. 1-30.
89. O. A. Awofala, Adeneye, and Abisola O. Lawani. 2020. "Increasing Mathematics Achievement of Senior Secondary School Students through Differentiated Instruction." *Journal of Educational Sciences* 4 (1): 1. https://doi.org/10.31258/jes.4.1.p.1-19.

90. Ojose, Bobby. 2008. “Applying Piaget’s theory of cognitive development to mathematics instruction.” *The mathematics educator* 18.1.
91. Olivia, P. F. 1997. Developing the curriculum (4th Ed.). New York, NY: Longman
92. Oluwatoyin, Adeleke Joshua. 2007. “GENDER ROLE ON TOPIC PREFERENCE OF SENIOR SECONDARY SCHOOL STUDENTS IN MATHEIVIATI CS,” 14.
93. Ornstein, A. C., &Hunkins, F. P. 2009. Curriculum foundations, principles and issues (6th ed.). Boston, MA: Pearson Education
94. Panthi, Ram Krishna, and Shashidhar Belbase. 2017. “Teaching and Learning Issues in Mathematics in the Context of Nepal.” Preprint. other. https://doi.org/10.20944/preprints201706.0029.v1.
95. Pech, P. 2010. On Proving and Discovering Theorems by Computer in Teacher Training. In T. Bianco, & V. Ulm (Eds.), Mathematics Education with Technology-Experiences in Europe, (pp. 230-243). Augsburg: University of Augsburg.
96. Pesen, C. Yapılandırmacı Öğrenme Yaklaşımına Göre Matematik Öğretimi. Ankara: Pegem, Turkey, 2008.
97. Pia, Kaniz Fatema. 2015. “Barriers in Teaching Learning Process of Mathematics at Secondary Level: A Quest for Quality Improvement.” *American Journal of Educational Research* 3 (7): 822–31.
98. Piaget, Jean. 2000. “Piaget’s theory of cognitive development.” *Childhood cognitive development: The essential readings* 2: 33-47.
99. Pietsch, James, Richard Walker, and Elaine Chapman. 2003. “The Relationship among Self-Concept, Self-Efficacy, and Performance in Mathematics during Secondary School.” *Journal of Educational Psychology* 95 (3): 589–603. https://doi.org/10.1037/0022-0663.95.3.589.
100. Polydoros, Georgios. 2021. “Teaching and Learning Mathematics with Mobile Devices,” 8.
101. Porter, Andrew. 1989. “A Curriculum out of Balance,” 7.
102. Rajkumar, R. & Hema, G. 2016. Modern mathematics classrooms: facilitating innovative teaching methods and learning strategies for 21st century learners. Edusearch, 7, 70-74.
103. Rampal, A., Ramanujam, R. & Saraswati, L. S. 1998. Numeracy counts! Mussoorie: National Literacy Resource Centre.
104. Register, Jordan, Michelle Stephan, and David Pugalee. 2021. “Ethical Reasoning in Mathematics: New Directions for Didactics in U.S. Mathematics Education.” *Mathematics* 9 (8): 799. https://doi.org/10.3390/math9080799.

105. Romberg, T. A. 2001. Standards-based mathematics assessment in middle school: Rethinking classroom practice. New York: Teachers College Press.
106. Romberg, Thomas A. 1999. *Mathematics Classrooms That Promote Understanding*. Routledge.
107. Sa'ad, Tata Umar, Abba Adamu, and Abdullahi M Sadiq. 2014. "The Causes of Poor Performance in Mathematics among Public Senior Secondary School Students in Azare Metropolis of Bauchi State, Nigeria." *Journal of Research & Method in Education* 4 (6): 32.
108. Sahlberg, P. 2006. Models of curriculum development: International trends and the way forward. In Sahlberg P. (Ed.), Curriculum reform and implementation in the 21st century: Policies, perspectives and implementation (pp. 108 – 121). Proceedings of the International Conference on Curriculum Reform and Implementation in the 21st Century.
109. Samuelsson, Marcus, and Joakim Samuelsson. 2016. "Gender Differences in Boys' and Girls' Perception of Teaching and Learning Mathematics." *Open Review of Educational Research* 3 (1): 18–34. https://doi.org/10.1080/23265507.2015.1127770.
110. Saxena, Ritu, Keerty Shrivastava, and Ramakant Bhardwaj. 2016. "Teaching Mathematical Modeling in Mathematics Education." *Journal of Education and Practice* 7.11: 34-44.
111. Schleppegrell, Mary J. 2007. "The Linguistic Challenges of Mathematics Teaching and Learning: A Research Review." *Reading & Writing Quarterly* 23 (2): 139–59.
112. Schlöglmann, Wolfgang. 2008. "Is Cognitive Neuroscience Relevant to Mathematics Education Research?," 5.
113. Schulze, Salomé, and Anne Bosman. 2018. "Learning Style Preferences and Mathematics Achievement of Secondary School Learners." *South African Journal of Education* 38 (1): 1–8.
114. Scriven, M. 1967. The methodology of evaluation, in: R. Tyler, R. Gagné& M. Scriven (Eds.), Perspectives of Curriculum Evaluation (AERA Monograph Series on Curriculum Evaluation), 1, 39-83.
115. Shumway, Richard J. 1980. "Research in Mathematics Education."
116. Sidhu, K. S. 1995. The Teaching of Mathematics. New Delhi: Sterling Publishers Pvt.Ltd.
117. Siegler, Robert S. 2000. "The rebirth of children's learning." *Child development* 71.1: 26-35.
118. Silver, Edward A, and Jeremy Kilpatrick. 1994. "E Pluribus Unum: Challenges of Diversity in the Future of Mathematics Education

Research." *Journal for Research in Mathematics Education* 25 (6): 734–54.

119. Skovsmose, Ole. "Mathematical education and democracy." *Educational studies in mathematics* 21.2 1990: 109-128.
120. Smedt, Bert, and Lieven Verschaffel. 2010. "Traveling down the Road: From Cognitive Neuroscience to Mathematics Education … and Back." *ZDM* 42 (6): 649–54. https://doi.org/10.1007/s11858-010-0282-5.
121. Smith, John P. 1996. "Efficacy and Teaching Mathematics by Telling: A Challenge for Reform." *Journal for Research in Mathematics Education* 27 (4): 387–402.
122. Sobel, D.M., 2010. Integrating top-down and bottom-up approaches to children's causal inference. In: Johnson, S.P. (Ed.), Neoconstructivism: The New Sciences of Cognitive Development. Oxford University Press, Oxford, pp. 159–179.
123. Stewart, Sepideh, ed. 2017. *And the Rest Is Just Algebra*. Cham: Springer International Publishing. https://doi.org/10.1007/978-3-319-45053-7.
124. Sullivan, Peter, Steve Tobias, and Andrea McDonough. 2006. "Perhaps the Decision of Some Students Not to Engage in Learning Mathematics in School Is Deliberate." *Educational Studies in Mathematics* 62 (1): 81–99. https://doi.org/10.1007/s10649-006-1348-8.
125. Sutherland, Rosamund, Victoria Armstrong, Sally Barnes, Richard Brawn, N Breeze, M Gall, Sasha Matthewman, et al. 2004. "Transforming Teaching and Learning: Embedding ICT into Everyday Classroom Practices." *Journal of Computer Assisted Learning* 20 (6): 413–25.
126. Taba, H. 1962. Curriculum development: Theory and practice. New York, NY: Harcourt, Brace & World Inc.
127. Tang, Akaysha C., et al. 2006. "Programming social, cognitive, and neuroendocrine development by early exposure to novelty." *Proceedings of the National Academy of Sciences* 103.42: 15716-15721.
128. (Tasos) Barkatsas, Anastasios, and John Malone. 2005. "A Typology of Mathematics Teachers' Beliefs about Teaching and Learning Mathematics and Instructional Practices." *Mathematics Education Research Journal* 17 (2): 69–90. https://doi.org/10.1007/BF03217416.
129. Teinken, C. H. 2013. Conclusions from PISA and TIMSS testing. Kappa Delta Pi Record, 49, 56–58.

130. Thanheiser, Eva, Christine A Browning, Meg Moss, Tad Watanabe, and Gina Garza-Kling. 2010. "Developing Mathematical Content Knowledge for Teaching Elementary School Mathematics." *Issues in the Undergraduate Mathematics Preparation of School Teachers* 1.
131. Turgut, İlknur Gülşen, and Sedat Turgut. 2020. "Me While I Am Learning Mathematics: Reflections to Elementary School Students Drawings." *Lnternational Electronic Journal of Elementary Education* 13 (1): 139–54. https://doi.org/10.26822/iejee.2020.179.
132. Tyler, R. 1949. Basic principles of curriculum and instruction. Chicago, IL: The University of Chicago Press.
133. Umameh, Michael. 2011. "A SURVEY OF FACTORS RESPONSIBLE FOR STUDENTS POOR PERFORMANCE IN MATHEMATICS IN SENIOR SECONDARY SCHOOL CERTIFICATE EXAMINATION (SSCE) IN IDAH LOCAL GOVERNMENT AREA OF KOGI STATE, NIGERIA." https://doi.org/10.13140/RG.2.1.1965.7766.
134. University Grants Commission. 2011. 11th plan guidelines. Retrieved April 9, 2012 from http://www.ugc.ac.in/fi nancialsupport/xiplan/guideline.htm.
135. University Grants Commission (n.d). Model Curriculum.Retrieved April 9, 2012 from University Grants Commission Website: http://www.ugc.ac.in/policy/modelcurr.
136. Van den Heuvel-Panhuizen, M. 2003. The didactical use of models in Realistic Mathematics Education: An example from a longitudinal trajectory on percentage. Educational Studies in Mathematics, 54, 9-35.
137. Van den Heuvel-Panhuizen, M. H. A. M. *Assessment and realistic mathematics education*. Vol. 19. Utrecht University, 1996.
138. Van den Heuvel-Panhuizen, Marja, and Paul Drijvers. 2020. "Realistic mathematics education." *Encyclopedia of mathematics education*: 713-717.
139. Varma, Sashank, Bruce D. McCandliss, and Daniel L. Schwartz. 2008. "Scientific and pragmatic challenges for bridging education and neuroscience." *Educational researcher* 37.3: 140-152.
140. Vosniadou, S., Vamvakoussi, X., Skopeliti, I., 2008. The framework theory approach to the problem of conceptual change. In: Vosniadou, S. (Ed.), International Handbook of Research on Conceptual Change. Routledge, New York, pp. 3–34.
141. Wardlaw, C. 2008. Presentation to the National Curriculum Forum held in Melbourne, 2008.

142. Wellman, H.M., Lagattuta, K.H., 2004. Theory of mind for learning and teaching: the nature and role of explanation. Cognitive Development 19, 479–497
143. Wigley, A. "Models for Teaching Mathematics" ATM-The Association of Teachers of Mathematics, 2008.
144. Wiles, J. 2008.Leading curriculum development. London, United Kingdom: Sage Publications.
145. Wilson, W Stephen. 2009. "Elementary School Mathematics Priorities," 10.
146. Yetkin, Elif. 2003. "Student Difficulties in Learning Elementary Mathematics. ERIC Digest."
147. Zakaria, Effandi, Titi Solfitri, Yusoff Daud, and Zulkarnain Zainal Abidin. 2013. "Effect of Cooperative Learning on Secondary School Students' Mathematics Achievement." *Creative Education* 04 (02): 98–100. https://doi.org/10.4236/ce.2013.42014.
148. Zelazo, P.D., Müller, U., Frye, D., Marcovitch, S., 2003. The development of executive function in early childhood. Monographs of the Society for Research in Child Development 68 (3) i-151.
149. Zevenbergen, R. 2000. "The Construction of Identity in Secondary Mathematics Education," 14.